The National Computing Centre develops techniques and provides aids for the more effective use of computers. NCC is a non-profit-distributing organisation backed by government and industry. The Centre

- co-operates with, and co-ordinates the work of, members and other organisations concerned with computers and their use
- provides information, advice and training
- supplies software packages
- promotes standards and codes of practice

Any interested company, organisation or individual can benefit from the work of the Centre by subscribing as a member. Throughout the country, facilities are provided for members to participate in working parties, study groups and discussions, and to influence NCC policy. A regular journal – 'NCC Interface' – keeps members informed of new developments and NCC activities. Special facilities are offered for courses, training material, publications and software packages.

For further details get in touch with the Centre at Oxford Road, Manchester M1 7ED (telephone 061–228 6333)

or at one of the following regional offices:

Belfast
1st Floor
117 Lisburn Road
BT9 7BP
Telephone: 0232 665997

Birmingham
2nd Floor
Prudential Buildings
Colmore Row
B3 2PL
Telephone: 021–236 6283

Bristol
6th Floor
Royal Exchange Building
41 Corn Street
BS1 1HG
Telephone: 0272 27077

Glasgow
2nd Floor
Anderston House
Argyle Street
G2 8LR
Telephone: 041–204 1101

London
11 New Fetter Lane
EC4A 1PU
Telephone: 01–353 4875

Systems Documentation

S. J. WATERS

Keywords for information retrieval (drawn from the *NCC Thesaurus of Computing Terms*):
Systems Documentation, Languages

British Library Cataloguing in Publication Data

Waters, Samuel Joseph
Systems specifications. —
(The LSE/NCC monographs on information systems analysis and design).
1. Information storage and retrieval systems
1. Title II. Series
001.5 T58.6

ISBN 0-85012-188-4

Published in April 1979 by:
NCC Publications, The National Computing Centre Limited,
Oxford Road, Manchester M1 7ED

Printed in England by H. Charlesworth & Co Ltd,
Huddersfield

ISBN 0-85012-188-4

Acknowledgements

The author wishes to acknowledge:

— London University colleagues, particularly Bob Davenport, Sandy Douglas, John Florentin, Kit Grindley, Jim Inglis, Peter King, Frank Land and Ronald Stamper, for their constructive criticisms and comments.

— The Science Research Council, for financing some of the research towards the book.

— M and Mme Bouyala, for providing the perfect atmosphere for writing at Le Moulin Neuf, Uzès in the Erôle region of Southern France.

— Susan Coles, for patiently interpreting the manuscript.

This book is dedicated to the memory of the late T R Thompson, the man who, above all others, perceived the impact of computers in business during the early days of LEO.

S.J. Waters

Preface

Systems documentation, one of the most important aspects of Informatics, involves recording the progress and results of one's work for the benefit of others.

An organisation faced with a typical information processing problem needs to draw on the skills of various people:

— users — management, white-collar and blue-collar workers;

— systems analysts — to solve the business problems (business systems analysts) and to specify this solution as an outline man/machine system (computer systems analysts);

— systems designers — to detail the manual and computer systems;

— programmers — to code the computer system;

— operators — to prepare computer input and run the computer system.

Sometimes these various jobs are allocated to different teams but one person may be both a user and a systems analyst or both a systems analyst and designer. Other specialists are often involved, such as O & M and work-study officers, ergonomists, sociologists, accountants and auditors, mathematicians (particularly statisticians and operational researchers), computer manufacturers, telecommunication experts, stationers, and so on.

Clearly, good documentation is vital if all these people are to communicate successfully with each other. It is also vital within each job as one team member may be called upon to take over the work of another.

The most important document produced during a systems project is usually the specification of user requirements, sometimes termed a *systems spec, functional spec,* or *proposal.* This document is produced by systems analysts to record the information processing requirements of user. After user agreement it is passed to systems designers. The specification is analogous to an architect's scale drawing of a new building: the buyer agrees the drawing before it is passed on for detailed design and construction. When the building has been constructed, such scale drawings are retained to permit extensions and revision. Similarly, a specification continues to be a vital document when the system is operational.

This book is concerned solely with such documentation. It attempts to define the facts that may be included in a specification of user requirements, to discuss some standard documentation methods used to record these facts, and to survey some current research and development into higher-level systems languages.

This book recognises that the prime function of a specification is to contain the facts from which the system can be designed and constructed. Thus we analyse the decisions taken by a systems designer and from these deduce which facts are necessary for the required decisions. Practical examples illustrate the importance of these facts.

It is concluded that:

- — probably no existing standard documentation method nor systems language has the capacity to record all relevant facts for initial systems design or subsequent modification to that design;
- — it is probably not feasible to document all the facts that might be significant to a designer.

There is no convenient division between systems analysis and systems design: the analysis team cannot completely finish its specification work before the design team starts any of its construction work. This can pose severe problems for organisations which segregate systems analysis and design.

This book is mainly aimed at trainees and students, but experienced systems analysts and designers may also find it useful. Many of the points discussed are as relevant to manual and electromechanical systems as they are to computer systems. Thus the book is not restricted to computer specialists.

Contents

1 Introduction

GENERAL

A jeweller developed a complex production explosion system to calculate the quantities of raw materials, component parts and sub-assemblies needed to manufacture given quantities of finished products. The system was abandoned after two years because it could not cope with more than 3,000 products.

A car manufacturer organised its product/sub-assembly/component part/raw material master records as a magnetic disk, indexed-sequential file. Within a couple of months running, the overflow patterns made the file so inefficient that the system was reprogrammed to organise the records as a serial, magnetic-tape file.

A mail-order company installed an accounting system to control housewives' books. An unexpectedly large team of clerks had to be employed to correct the high proportion of errors that the computer found in payments data.

A consultant engineering company was about to abandon its computer-based payroll and job-costing system because it was inflexible. At the eleventh hour, it was realised that their engineers could be paid overtime an extra week in arrears which solved the problem and kept the system operational.

All these experiences are true and similar to those of many organisations using computer-based systems. All the organisations made the common mistake of omitting facts from the specification of the system: the jeweller did not state the maximum number of products; the car manufacturer did not state the point-overflow pattern on the master file; the mail-order company did not state the accuracy of its raw-data from housewives; and the consultant engineering company nearly did not state the time delay between notifying and processing an overtime claim. What is the point of a brilliantly designed and programmed system which fails because important assumptions are false?

Hundreds of systems design and programming methods and case studies have been analysed to define the facts that may be recorded in a specification. An interesting theoretical question is whether a complete specification method, containing all facts, could ever be developed.

The term *facts* includes both facts *per se* and estimated facts and opinions. Much of the literature on fact-finding discusses how to prepare, conduct and follow-up interviews and how to organise surveys using questionnaires but fails to indicate what facts may need to be found. This book hopes to redress the imbalance and does not concern itself with fact-finding methods.

Systems analysts write specifications incorporating user requirements. They state the procedures that must be programmed into a computer system, plus the information that must be input and output by these procedures. Typical specifications run to hundreds of pages.

Once a specification has been agreed by the users, it is passed to systems designers who decide how the computer can achieve the user requirements. This involves designing a computer system of files and programs, evaluating it (eg, estimating computer time and costs) and specifying the design for subsequent programming.

The specification of a computer system that is agreed by the user and passed from a systems analyst to a systems designer is a vital link in the communication chain between users, analysts, designers, programmers and operators. The specification is a vital document requiring user agreement. This book should help to overcome one of the most common mistakes in specifications, that of omitting facts that might be highly significant to the designer.

Some research teams, notably in the USA, are attempting to develop formal higher-level systems languages that completely specify user requirements. These requirements are then fed to computer software which checks consistency and assists the designer or, in some cases, attempts automatic design and programming of the computer system. If relevant design facts are omitted automatic design cannot be achieved.

SYSTEMS

In this book a *system* is an *information system,* information processing system, or data processing system. The parts of these systems are information (eg, hours worked, pay rate, basic pay) and the relationships are the procedures that transform *given* information into *derived* information (eg, the 'calculate pay procedure' transforms hours worked and pay rate into basic pay). The term *information* here refers loosely to everything on messages and records. At the lowest level is an *inform-*

ation element: elements may be grouped together into *information sets* (for example, employee payslip). Elements and sets can be grouped into larger sets, for example payroll information, to form a hierarchy.

Similarly, at the lowest level of processing is a *procedure* which derives an element (for example, calculate pay). Procedures may be grouped into systems (for example, produce payslip). Procedures and systems can be grouped into larger systems (for example, process payroll) to form another hierarchy.

The varieties of information systems discussed in this book are mainly those that are *computer-based.* These systems have some of their information represented as files on computer storage devices, such as magnetic tapes and disks. Computer programs are coded in various languages (mainly COBOL) to process these files.

Many attempts have been made to classify computer-based information systems and one early dichotomy was between *software systems* and *application systems.* Software systems, now usually termed *systems software,* were regarded as general-purpose, computer-housekeeping programs that could be used over a wide variety of computer applications and included computer operating systems, compilers, assemblers and utilities (eg, sort programs). *Application systems* were individually developed to solve a particular information processing problem in commercial or scientific/technical fields. Although the examples used here are drawn from the commercial field, the theories presented are often equally applicable to scientific/technical systems, systems software and any other type of computer-based information system.

Many attempts have also been made to classify computer systems. For example, a *real-time system* is one which controls an environment by receiving the input data, processing it and returning the output results sufficiently quickly to affect the functioning of the environment at that time. A *batch-processing system* is one in which the input data is collected into batches for periodic processing into the output results. Further classifications include quick-response systems, on-line systems, multi-access systems and remote job entry systems. In this book, although practical examples are drawn mainly from the field of batch-processing systems, the theories presented are often equally applicable to all types of computer system.

THE SYSTEM'S LIFE-CYCLE

When a computer-based information system of even moderate complexity has to be developed, experience has shown that it is wise to proceed through an orderly sequence of actions, sometimes referred to as the system's life-cycle. It has an iterative nature, any stage looping back to a previous state, proceeding to the next or leading to aborting the project.

Systems definition attempts to specify the requirements to be achieved by the total system. The result is a 'black-box' specification of the total system in terms of its input and output information and procedures. This specification work is usually carried out by systems analysts in close consultation with users, and is so costly in time and money that it is often preceded by an outline feasibility study to indicate if the effort is likely to be worthwhile.

Systems design first partitions the total system into two smaller black boxes, the human and computer systems. The communications between these systems and with their environment are then designed, specifying data capture methods, form layouts, teleprocessing networks, and so on. The human system is designed as clerical procedures, ledgers, etc, and the computer system as programs, computer files, etc. This design is usually carried out by systems designers who may also be systems analysts, senior programmers or database administrators. Often, it involves innovation, such as research into new computer hardware and software or into new data capture methods using terminals. Several designs may need to be compared.

For the human system, systems implementation includes training, documentation, and file creation, and for the computer system, flow-charting, coding and 'debugging'. Finally, the changeover procedures involve systems testing and pilot parallel running. Usually several man-years' effort and tens of thousands of pounds are spent in implementing even the simplest systems.

Many systems are risky because they incorporate innovations or are based on estimates so these are often implemented gradually. Thus the 'backbone' of the system is implemented first. After experience has been gained running the basic system, it is modified and extended to include the remainder of the design.

The live computer system is then run by the computer operations department. The total system should be monitored and periodically reviewed as part of systems maintenance. This may result in returning to square one to completely alter some parts of the system or to incorporate previously unforeseen requirements.

The systems designer must be informed of the resources that will be available and the objectives and constraints that should be met. Resources include the men, money and machines (for example, programmers, operators, computer hardware, computer software, etc), constraints being the periods for which these will be available. Objectives have a profound impact on the internal design of the computer system and include:

Efficiency — The computer system should not waste costly, and often scarce, computing resources, eg don't buy eight magnetic disk drives if two are ample.

Timeliness — The computer system should meet turnaround/ response times so that operational outputs are produced on schedule, eg employees must be paid on time.

Security — The computer system should recover from breakdowns so that it gives a regular, reliable service to the user, eg stockbrokers can go bankrupt if the system is off the air for a couple of days.

Accuracy — The computer system should work to an acceptable level of correctness. Inaccuracy can cause plane and train crashes, etc.

Compatibility — The computer system should fit in with existing and future computer systems to achieve an integrated whole, eg make sure the personnel records system 'talks to' the payroll system (via computer files).

Implementability — The computer system should be technically feasible and constructable with respect to available resources, eg don't expect trainee programmers to produce a data verification program (usually they fail and the entire system is delayed).

Maintainability — The computer system should be modifiable with respect to continuing resources.

Flexibility — The computer system should be able to cope with changes to its logic, eg governments are forever changing taxation laws.

Robustness — The computer system should be able to cope with changes to its statistics of throughput, traffic or workload.

Portability — The computer system should be able to cope with changes to its hardware/software configuration, eg you may be able to sell your payroll system to a similar organisation with different equipment.

Acceptability — The computer system should satisfy any design standards imposed by the organisation.

Economy — The computer system should be cost-effective or cost-beneficial when compared with its alternatives.

In practice because objectives conflict with each other and their relative importance varies for each computer system, the designer should be informed of the dominant objectives and their relative weightings. For example, a payroll system demands flexibility to keep up with changing statutes, whereas a hospital-patient monitoring system demands accuracy to save human life; possibly the payroll demands less accuracy and the hospital less flexibility. The choice for a bank, for example, may lie between:

— a centralised, daily-updating computer which is cheap;

— a centralised, immediate-updating computer which is accurate;

— a decentralised network of computers which is flexible.

Thus the resources and objectives are varied to achieve different designs for evaluation and comparison.

THE BLACK BOX

In theory the user of a computer system can view it as a black box that operates some information processing requirements subject to the system's objectives and constraints. It is necessary to master the methods of using the system effectively. In practice, those users without knowledge of computer technology can operate in this way, given good technical support.

The black box should be built to the user's specification of requirements and objectives translated into a computer system of files and programs by systems designers and programmers. Unfortunately some computer technocrats betray the user's trust by mistakenly placing their allegiance with the 'computer profession' instead of with the user they are supposed to serve. In a well-motivated organisation the user is served by the computer staff.

In one instance a user's systems analysts completely specified his requirements for a comprehensive payroll system. This specification was not altered at all during systems design, implementation and initial operation. The first section of the specification defined the objectives to be met by the computer system and clearly stated that computing efficiency was not dominant. Programmers designed a system based on low-level assembler language programming to save a few minutes computer time per week and make life more interesting for them! Estimated programming costs were halved by insisting that a high-level language be used, which also met other dominant objectives such as maintainability and portability. In this case, the user's interests were corrupted. The black box view can be dangerous in such an environment.

The black box represents the logical system and ignores the technical intricacies of the physical system. This enables the user to concentrate on what is done without being confused by how it is done. It matters little whether the physical system is computerised or manual.

There are three essential parts to any computer system. First, the system must produce output messages (such as payslips). Second, it must be activated by input messages (such as clock cards). Third, the system must transform the input messages into output messages. Thus every computer system consists of input messages, procedures and output messages.

In particular, the input messages necessary to activate an output message often include some current information such as hours worked, and a mass of historic information such as payrate and pay for the year to date. This information may be stored by the computer system as a database of computer files to avoid continually re-inputting it from the human system. Thus in accordance with current practice it will be regarded as a necessary fourth part.

Finally, because many information elements are input, stored in the database, and output, it is inconvenient to completely define their characteristics every time they are mentioned. Instead, a data dictionary can relate information characteristics by centralising their definition into a composite part which provides a backbone for system definition. Thus the user's requirements may be specified in terms of five parts: data dictionary, input messages, database, output messages and procedures.

SPECIFICATIONS

A specification defines the black box required by users in terms of at least the data dictionary, messages, database and procedures; objectives, and possibly available resources, are also included. Factual completeness requires a potentially vast amount of information to be established, analysed and recorded. For example, a typical payroll specification might require one year's work by a systems analyst. Subsequently the systems designer finds that facts are missing and raises 'systems queries' which lead to further analysis work. These queries are raised by designers and are dictated by the technical alternatives they consider. Enthusiastic designers may raise many different queries because they explore more alternatives. In theory the payroll specification could easily run to thousands of pages if all possible design facts were recorded.

How then do we cope in practice? Specification methods have evolved over the past twenty years to include facts that are essential whatever design approach is taken and those that are probable (ie likely to be useful to the systems designer). Facts which are only useful occasionally are deliberately omitted to save specification effort in the hope that few systems queries will arise. The following seven chapters contend that these practices, based on human judgement and past experience, are so risky that they justify critical examination. For example, most specifications omit essential as well as probable facts, and many of the possible and recorded facts are often useless to the designer. Thus many systems analysts are wasting time and effort.

Many systems designers mistakenly accept the specification as a complete statement of the problem to be solved, failing to recognise:

— The semantic problems in organisations. What is a customer report? Is it a printout of delivery points, including branches of chain stores, for the transport manager? Is it a printout of sales points, including area buying offices of chain stores, for the sales manager? Or is it a printout of billing points, including the head offices of chain stores, for the chief accountant? What is a year calendar, financial or tax? The wise designer will ensure that systems analysts check that the specification is correct. For example, a UK Cancer Research Group specified a system for maintaining a large computer file of all cases, whether alive or dead, with details of their treatments and the results. One designer assumed 'patients' were cases and quoted a solution, costing over £30,000 per annum, to update the entire file every month. A wiser designer quoted an alternative solution, costing less than £10,000 in which the dead cases were not regularly updated for treatment results.

— The specification is not complete. The lazy designer assumes it is and chooses design options by what is recorded in the specification, often failing to investigate possibilities beyond the first successful one. A keen designer will explore more options and demand more facts from the systems analysts, and will often achieve major savings as a result. For example, a large payroll system used magnetic tapes to store the employee file in building site/employee number sequence. The specification included an amendment form to alter the building site for employees who change their workplace. One designer suggested a system which processed the entire file twice every week; another realised that these amendment forms could easily be submitted one week prior to the changes taking place. This halved the file processing time and saved hours of computer time every week, saving thousands of pounds per annum.

— The specification is not a definitive statement of what the user wants and can afford. It may record what the user wants ideally but the costs are often open to negotiation. Thus there is often room for manoeuvre when balancing user requirements and their costs.

Many organisations have combined their payroll and personnel systems to use the same database. This has the important advantage of achieving accuracy in the personnel system by ensuring that each employee is recorded once only. However, the problem has often arisen of the payroll system updating the same record that the personnel system is trying to update, thus the database becomes shared and the solutions are expensive (often requiring costly database management software). The designer should try to avoid this situation.

One solution is to accept that employee personnel information can be updated at any time except for the couple of hours per week when the payroll system updates this information. This relatively minor inconvenience can save considerable sums and is basically equivalent to personnel and wages clerks having slightly restricted access to shared employee records to avoid them getting in each other's way.

Designers must not be allowed to disregard the meaning, completeness and accuracy of the specification. After all, it is simply a communication between colleagues — systems analysts working with people and systems designers working with machines. The communication must be 'two-way' to achieve better end-results for the user who has to pay for the consequences.

Rivalry still exists between analysts and designers, and a strong argument exists for having one person doing both jobs. If the analysis and design jobs are combined, the systems specification is still a vital document: it is the blueprint for further design and construction work that must be agreed by the user.

The specification should contain all the facts which are relevant to the designer and these should be perfectly accurate and cleansed of all semantic problems. The following chapters suggest a classification of design facts that should improve current specification methods. Each fact specified in the data dictionary, messages, database and procedures is first classified as *logical* or *statistical,* and then subclassified as *necessary, probable* or *possible.* Figure 1.1 outlines this scheme.

Systems logic, often termed systems algebra, contains all the facts that contribute to the essential working of a system and which are typically coded into computer programs. Some logical facts are absolutely necessary to build a working system, irrespective of the computer resources or design strategies employed. Conceptually, these facts would be necessary even if a super-computer were used which cost nothing, took no time to perform its operations and had infinitely large storage capacity. Other logical facts may be probably or possibly relevant to the designer depending on whether these are usually or only occasionally required.

Systems statistics, often termed systems arithmetic, contain all the facts that allow designers to evaluate the performance of a system in terms of its times and costs. These numerical facts include traffic, workload or throughput (eg, number of messages per hour), data volumes (eg, number of records on file) and distributions (eg, proportions of records that are updated each day). Similar to logical facts, statistical

	Systems Logic			Systems Statistics	
	Necessary	Probable	Possible	Probable	Possible
Data Dictionary					
Messages					
Database					
Procedures					

Figure 1.1 Classification scheme for specification facts

facts are probably useful to the designer whereas others are only possibles.

Some organisations take the view that estimating is a waste of time because current computer technology makes it too difficult and there are too many imponderables; but often they play into the hands of computer manufacturers by buying themselves out of trouble when it is subsequently found that the daily programs take thirty hours to run or that the main files are too large for available storage. At the other extreme, some systems are estimated to the nth degree by building prototypes and simulators. Often the results bear little resemblance to reality because they may be based on only first degree approximations (eg, what in fact, will the number of personnel enquiries be per hour?). Between these extremes, every system must be estimated to a reasonable extent.

The following chapters build on this classification scheme to suggest which facts might need to be specified and whether they are necessary and must be specified or are probables, and should be established by systems analysts. If they are only possibles they are best left to designers subsequently raising systems queries. An organisation can compare its own specification methods against this scheme.

DOCUMENTATION STANDARDS

Numerous documentation standards have evolved over the last two decades to improve communication between analysts and designers by formally stating which facts must be recorded in a systems specification. Usually, these standards include preprinted forms to guide the analyst. Chapter 9 discusses three such documentation standards and compares them against the previous classification scheme to show that none is complete.

SYSTEMS LANGUAGES

Gradually, analysts and designers are realising that they should possibly take advantage of the computer to eliminate some of their own manual effort. If the computer is an ideal number-crunching tool for the payroll department and an ideal information-retrieval tool for the sales department then why object to its use in the systems analysis and design department?

We should view the system's life-cycle as an information processing problem in order to develop a computer-aided methodology. Some research and development teams, particularly at Scandinavian and US universities, are attempting this approach and are suggesting that higher-level systems languages could be used to formally document specifications which can then be processed by computer software.

Information Algebra was developed in the early 1960s by the US CODASYL Development Committee which had previously given us COBOL. The language used mathematical notations which confounded many people and helped to bury its concepts. Systematics was suggested in the mid 1960s by Dr Kit Grindley of Urwick Dynamics Ltd, and after continuing development, it is now receiving much interest in the UK and Europe. PSL (Problem Statement Language) was produced by ISDOS, the Information System Design and Optimisation System group at Michigan University, USA during the late 1960s. This pioneering group has received much world-wide attention, particularly for PSL which is one of the most complete and detailed attempts at a systems language actually being used in organisations. A search through the computing literature reveals dozens of systems languages.

CONCLUSION

The main objective is simply to help improve the current methods for specifying user requirements which have evolved from practice. This book provides a theoretical approach which justifies the facts recorded in specifications whereas a designer's approach can contain many omissions.

Chapters 2 and 3 suggest omissions at the black box level where a system is dissected into its parts: input and output messages, database and procedures. In general, six subsystems are identified which incorporate different types of procedure, to read six types of input message, and write six types of output message.

Chapters 4, 5, 6 and 7 suggest omissions at the detailed level of facts recorded for data dictionaries, messages, database and procedures, respectively. These facts are classified as necessary, probable or possible to indicate whether a systems designer must know them, usually needs to know them or only sometimes requires them (in which case a systems query is raised). Detailed checklists are developed as guidelines for specification methods and Chapter 8 suggests techniques for checking that a specification is reasonably complete and self-consistent.

No formal distinction is drawn between a documentation method and a language because in general, they may be viewed as the same thing. However, SPEC, ADS and the NCC Systems Documentation Standard seem to be mainly concerned with man-man communications whereas Information Algebra, Systematics and PSL also lend themselves to man-machine communications.

It is concluded that:

— there is probably a long way to go before any specification method, documentation standard or systems language is complete;

— it is probably not feasible to completely specify a system anyway;

— systems queries will continue, whether raised by human or automatic designers;

— there is no convenient point at which communications between analysts and designers can be completely severed.

A feasibility specification summarises the 'main' user requirements so they may be roughly designed to decide whether it seems worthwhile continuing with a detailed systems study. The subsequent detailed specification is the basis for a full design. Often the feasibility and detailed designs are completely different in strategy because seemingly trivial requirements can have far-reaching implications.

2 Anatomy of a system

INTRODUCTION

A decade ago many computers were actually ordered on the basis of limited knowledge in order to beat delivery delays. Also, viable computer-based systems have been rejected in the absence of more detailed analysis. The necessity for a detailed systems survey can thus help prove the case for using a computer and define which type of equipment is suitable; Alternatively, it can discard computers but point to improved manual or electro-mechanical systems.

It is essential to identify all parts of the system before any detailed fact-finding (eg defining the maximum number of products manufactured). First we decide the input and output messages, the database records, and the procedures necessary in the system and then we define their contents in detail.

The aim of this chapter is to answer the question 'What is a system?' before continuing in any detail, and it generalises the first level of systems definition.

Figure 2.1 illustrates the basic anatomy of an operational live system as six subsystems reading six types of input messages and writing six types of output messages. Figure 2.1 is a useful anatomy of most (although not definitive of all) systems which the remainder of this book explores in detail.

THE APPLICATION SUBSYSTEM

This contains all the parts of the system peculiar to the application and which form the basic user requirements. Application procedures process transactions against the application database which is updated to produce results as follows.

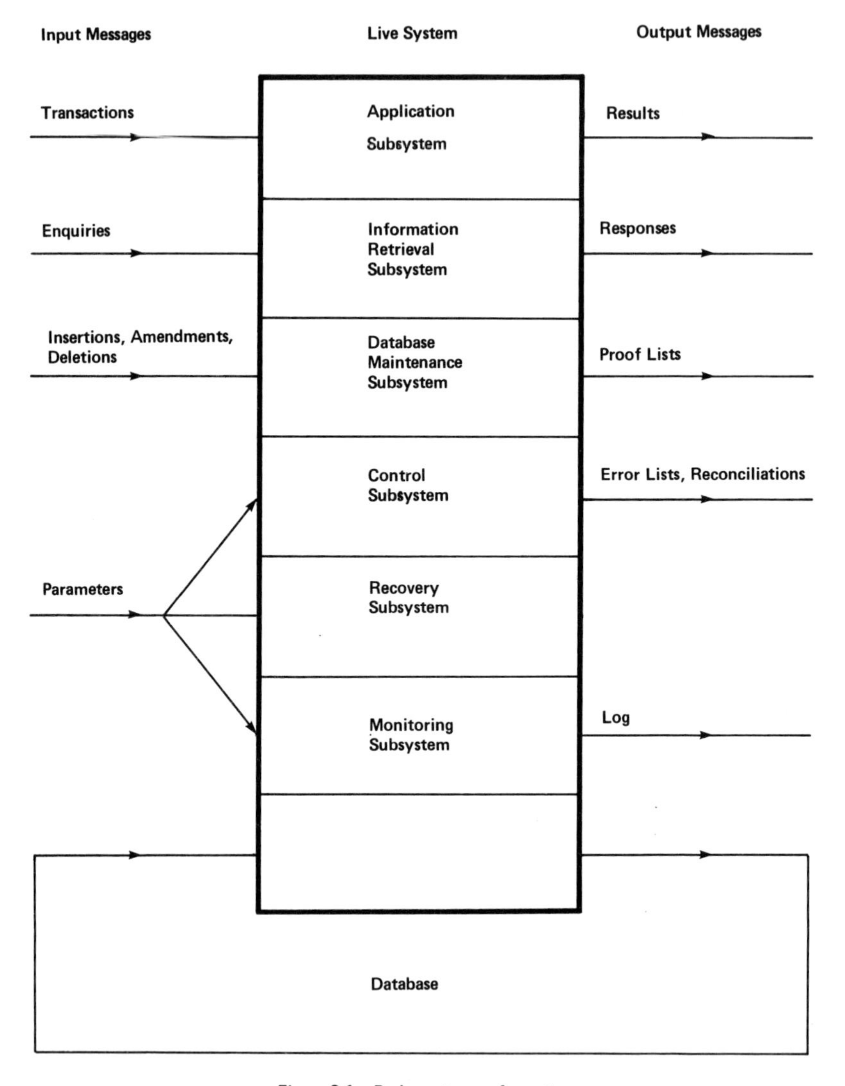

Figure 2.1 Basic anatomy of a system

Results

In the widest sense, results may be regarded as all output messages plus the updated version of the database. In this book, a narrower view is taken and results are the basic output messages from a system.

Transactions

Transactions are those input messages that record events which trigger results as output messages. Usually transactions are processed against the application database which is updated and results are generated. Generally, m different transactions will trigger n different results (where m and n are positive integers).

In a typical case:

— One delivery note and one invoice are normally triggered immediately by one customer order, but there may be a week's delay if only special products are requested. Stock-outs will sometimes trigger many delivery notes and invoices from an order because it can be satisfied by many part deliveries.

— One statement line is triggered at the end of each month for one sales ledger credit or debit input; another is triggered by an invoice, actioned by a customer order. Complete statements are triggered by an 'end of the month' parameter which is operated on the last Saturday of the month.

— One stock report line is triggered by many orders 'running' on a particular product.

— All stock forecast lines are triggered by a single 'end of the week' parameter.

— One credit or bad debt report can be triggered by numerous orders, sales ledger credits and debits.

Thus all results can be traced back to their triggering transactions to illustrate all 'm/n' relationships.

Application Database

The application database supports the transformations of transactions into results and must contain records of historic information necessary to these transformations. Historic is meant to imply that the information has previously been input to the system, although it may refer to long-past, recent or even future events. The database can only know about historic messages that were input in the past.

The application database must contain:

— Product, customer, outstanding order, discount and tax records to convert customer orders into delivery notes and invoices.

— Sales ledger records to produce monthly statements.

— Product records to produce stock reports and forecasts.

— Customer records to produce credit and bad debt reports.

Logically, a record is simply a convenient grouping of information elements. The customer records could be combined into a single record which might also include the sales ledger information and even the outstanding orders.

Application Procedures

Application procedures are the routines that are programmed to transform transactions into results and to update the application database. Clearly the user is aware of these procedures which in LEO's case include:

1 Stock allocation, invoice pricing, customer discounting and tax calculation to transform orders into delivery notes and invoices.

2 Sales ledger accounting and cash discounting to produce monthly statements.

3 Stock accounting and exponential smoothing to produce stock reports and forecasts.

4 Credit accounting and FIFO cash settlement to produce credit and bad debt reports.

The documentation of these procedures is likely to be a major task and many decisions will be made affecting such things as credit limits and customer discounts.

These procedures will eventually be supplemented by numerous physical application procedures which are invisible to the user. These include coding and decoding external formats to internal formats when writing and reading messages, file processing routines which match input messages with the database, sorting messages and records into prescribed sequences and numerous movements of information between storage devices

This application subsystem is sometimes termed the transaction processing subsystem, particularly in real-time working.

THE INFORMATION RETRIEVAL SUBSYSTEM

Some form of computer database is usually necessary to support the previous application subsystem and it can be stored on media which can only be accessed via the computer itself. Information retrieval procedures must be provided to input enquiries which request that database information be read and output as responses to these enquiries.

Responses

Responses contain information requested from the database. It is usually wise to allow any database information to be output via responses. Systems analysts and designers will need the facility to verify the database contents in times of suspected failure.

Enquiries

Enquiries are those input messages that generate responses as output messages by extracting information from, but not altering, the database.

Generally, m different enquiries will trigger n different responses. A typical case illustrates all 'm/n' relationships as follows:

— '1/1' is commonly called a *key-retrieval* enquiry because it defines the key number of the particular database record that is requested.

— '1/n' is commonly called a *content-retrieval* enquiry because it defines the profile of necessary conditions that requested records must satisfy.

— 'm/1' is commonly called a *summary* enquiry because it summarises records that satisfy a content-retrieval enquiry.

These examples trigger responses when the enquiries are input. A good example of delayed responses is provided by information dissemination systems where a researcher predefines his profile of literature interests and the system responds later when new books or papers are published.

Often the value of the resulting response to an enquiry may be less than the value of the costly processing which is generated. Thus it can be good practice to install a conversation between the enquirer and the system, which will indicate the likely quantity and cost of producing the final response together with the response time. For example a personnel manager might be drawing up a short-list of candidates for employment. If the manager submits the tentative job profile as a content retrieval enquiry, the intermediate response may indicate that the short-list will be too large; then the manager reframes the profile and tries again.

Information Retrieval Database

Generally, the information retrieval database supports all sub-systems, if all-embracing interrogation facilities are provided. It may also contain information particular to the users' requirements. Sales analyses could require the following:

— market area/product group — total gross and nett value of sales for each month this year and last year;

— salesman — total nett value of sales for each month this year;

— customer — total nett value of sales for this month and each of the three previous months;

— product — total quantity, gross value, nett value and profit of sales this year;

— customer — total gross value, nett value and profit of sales this year.

Information Retrieval Procedures

Information retrieval procedures are very similar to the application procedures, but they do not usually update the database. The decisions include matching records with profiles and the actions include summation, counting and so on. The user is aware of these logical procedures but not of the physical housekeeping procedures. Procedures are often included to restrict user access to a database that is shared by many users. These restrictions usually rely on passwords so that particular users may only interrogate information open to their own password.

This information retrieval subsystem can be a major part of the whole, particularly if full interrogation is allowed. Chapter 3 discusses some of the design considerations raised by ad hoc sales analyses and the wide choice for its special database contents.

Some terminal-based systems include an 'interactive lead through' facility whereby the user and computer embark upon a conversation to establish what a particular enquiry demands. For example, a hospital doctor might wish to find out the dose of insulin most recently given to a patient. Whatever the structure of an interactive conversation, each user entry can be regarded as an enquiry and each computer entry as a response. Thus one logical interrogation can consist of many physical interrogations.

THE DATABASE MAINTENANCE SUBSYSTEM

Computer database information can only be altered by the computer itself. Thus, database maintenance procedures must be provided to insert new records, amend the contents of existing records, and produce proof lists.

Proof Lists

Proof lists, or changes lists, display the alterations made by its maintenance subsystem. These are used by the human and environmental systems to prove the correctness of changes. Usually the proof list for insertions displays all information elements that have been set up including the old and new values of any amendments.

Insertions

Insertions are input messages that define the contents of new database records. If the insertion facility is used to create the initial database in addition to enhancing the subsequent live database, then usually all elements of all types of records must be set up.

Occasionally, records may be inserted before they actually come into effect, but a date or time is usually included for normal processing. Any orders before that date are rejected as errors.

Amendments

Amendments are input messages that notify changes to database elements. Usually, the new value of an information element is entered for the given record key number or an adjustment value is entered which changes the value of one or more elements within a database record. Sometimes, blanket amendments are useful to change information values of all database records satisfying given conditions. For example, 'promote all employees on grade 5 to grade 6'.

It is usually good practice to permit any database information to be amended, apart from record key numbers which might avoid subsequent design complications. This is particularly useful for error recovery in some situations, for example where a software bug has corrupted a few elements on a large file. These facilities however, may need to be restricted to particular users.

Some amendments can have delayed effect and include a date or time when they are to be activated. It is often advisable to notify an amendment to a record key number, if possible one run before the change is to take place.

Deletions

Deletions are input messages that destroy old database records. Usually they merely contain a message type and the key number of an obsolescent record; thus each deletion destroys a particular record. Sometimes blanket deletions are useful to destroy all records satisfying given conditions.

Deletions can sometimes suspend a record from normal processing until it is destroyed at a later date, if ever. For example, an employee who leaves an organisation has his payroll record deleted. However, the record may be retained on the database until the end of the tax year. Usually the normal deletion facility is also required to destroy records immediately, if only to remove those that have been incorrectly inserted.

Database Maintenance Procedures

These include all the physical housekeeping procedures. Usually the user's logical requirements are merely conditions governing the insertion and deletion of records. For example, if two new customers are notified with the same code number, which one should be accepted?

Sometimes ownership procedures are included to restrict the information that a user may alter in a database shared by many users. For example, a salesman may be allowed to adjust the discount percentages for one of his customers but no others. The passwords can be extended for ownership so that users may only insert, amend or delete certain information elements which are open to their own password. Usually, users will be able to interrogate more information than they can change because in order to write, they must be able to read.

Standard database maintenance packages are available but these subsystems are usually programmed anew and require a great deal of careful effort. Systems often fail because all insertion/deletion cases have not been considered.

THE CONTROL SUBSYSTEM

Mistakes are made by the people who provide the input messages. Machines also make mistakes when reading, transmitting, processing and writing information. Systems must include a control subsystem which help detect, locate and correct these mistakes and to produce error lists and reconciliations by using a control database and input parameters, as follows.

Error Lists

When errors are detected a message should be output to identify and rectify the mistake automatically. Some systems however, are too rigid in their control. For example the GIGO acronym for 'Garbage In — Garbage Out' does not consider the possibility of 'garbage' resulting from a corrupted database. Ideally, the entire system should be controlled, not just its input messages.

Reconciliations

The previous section noted the need for error lists where specific errors in information had been detected. Another control, at a more general level, is the reconciliation which attempts to equate overall totals computed during the operation of a system. These are equivalent to the control accounts of conventional accounting systems. If all is correct, the user is given some confidence in the overall accuracy of the system. If discrepancies occur, then the user is immediately warned that the system has failed in some respect.

Three types of reconciliation are common in practice, standard, current and running. a *standard* reconciliation attempts to equate an actual total summed during the operation of a system against a standard (or expected) total. The actual total should equal the standard total adjusted by any deviations from the standard. Figure 2.2. illustrates a standard reconciliation. A *current* reconciliation attempts to equate a

Total gross pay actually paid = + **Standard gross pay**
(being number of employees in payroll
x number of standard hours per working week
x standard hourly pay rate)

+ **Increments to standard gross pay**
(being overtime pay
+ bonus pay
+ advance holiday pay
+ etc)

— **Decrements to standard gross pay**
(being gross not paid re absence, sickness,
employees on holiday, etc)

Figure 2.2 Standard reconciliation covering all employees in a payroll system

Total adjustments posted to sales ledger = + **Total credits to sales ledger**
(being total payments
+ total credited returns
+ total credited breakages)

— **Total debits to sales ledger**
(being total invoice values)

Figure 2.3 Current reconciliation covering all customer accounts in a sales ledger system

total summed during the operation of a system against the totals of its components. Figure 2.3 illustrates a current reconciliation. A *running* reconciliation attempts to equate a total brought-forward from the previous operation of a system against the same total carried-forward to the next operation of the system. Figure 2.4 illustrates a running reconciliation. Notice that a running reconciliation usually requires the entire file to be processed if it is to cover all records.

These reconciliations only detect errors within one complete cycle of the system's operation. Error location can be made more precise by reconciling at convenient points during the run, possibly at breakpoints or even at a level of individual operation when a reconciliation fails. Alternatively, if this is not acceptable, running reconciliations on numbers of database records can indicate the number of records lost or gained, and running reconciliations on actual record key numbers can indicate which records have been lost or gained. Figure 2.5 illustrates these two reconciliations for a file of product records. The first discrepancy shows the number of records lost or gained and the second gives the sum of their key numbers. If only one record is involved, its actual key number is given. If several records are involved, there is a chance that they have consecutive key numbers and thus occur in one group of a sequential file. Dividing the given sum of their key numbers by their number indicates where the group is centred in the file.

In the system every database element could be covered by a running reconciliation. Too often, live systems do not detect common errors, but instead detect errors that will probably never arise.

Control Parameters

Control parameters are the input messages that trigger the control procedures into producing error lists and reconciliations. Usually, the user notifies information to control the run, such as the date or time, and to control the input messages, such as counts of their numbers or totals of their elements. Operators generate systems commands, often in the operating system's job control language, which define the identification numbers of the programs to be run, the magnetic tapes to be written on, the card reader to be used, and so on.

Control Database

Control information must be stored in the database to preserve its accuracy. Records will contain control elements, for example a check-total of all decimal information in LEO's product record. Files will contain control records, for example the brought-forward totals for running reconciliations.

Another candidate for the control database is a file of all the 'constants' that support the control procedures of the following section. These include the minimum and maximum value of every element (whether on database or messages) and its radix. These are often

+ Total outstanding balances brought-forward on sales ledger (being the carried-forward total from previous operation of the system)	= + Total outstanding balances carried forward on sales ledger
+ Total outstanding balances inserted into sales ledger (with new customers)	+ Total outstanding balances deleted from sales ledger (with old customers)
+ Total amendments to outstanding balances in sales ledger (new values – old values)	+ Total adjustments posted to sales ledger (from figure 2.3 reconciliation)

Figure 2.4 Running reconciliation covering all customer accounts in a sales ledger system

+ Total number of product records brought-forward	= + Total number of product records carried-forward
+ Total number of new product records inserted	+ Total number of old product records deleted

---

+ Total key numbers of product records brought forward	= + Total key numbers of product records carried-forward
+ Total key numbers of new product records inserted	+ Total key numbers of old product records deleted

Figure 2.5 Running reconciliations to assist error location in a file of product records

constants stored in the data definition of some database management schemes. In all cases, they are changed by the computer department when requested to do so. The users should be able to alter the control 'constants' directly, via the normal database maintenance system, without the assistance of the computer department.

Control Procedures

Users are aware of many control procedures, because they know the circumstances under which error lists are produced and how reconciliation totals are summed, but most are invisible to them including all the physical, housekeeping procedures.

Control procedures can check the accuracy of computations on information or of information itself. In the former, a calculation may be performed in different ways to yield a comparison check. For example, an air-traffic control system predicts flight-paths on three computers and then compares their results to determine majority agreement. Also, information on messages and records can be subjected to the following accuracy checks:

— *Format check* that elements are represented in the appropriate character set;

— *Radix check* that digits of numeric elements with non-decimal radix take values less than the radix;

— *Range check* that elements take values within predefined minima and maxima limits;

— *Check-digit verification* that a self-checking number is valid;

— *Consistency check* that several elements take consistent values if they are inter-related;

— *Hash-total check* that the sum of numeric elements on a message or record equals the entered, expected total;

— *Batch-total check* that the sum of a particular numeric element over a batch of messages equals the entered, expected total;

— *Existent code check* that the value of a coded element on a message actually exists;

— *Sequence check* that messages are presented in the appropriate order;

— *Completeness check* that all messages have been processed once only;

— *Comparison check* that duplicated messages agree.

These validation checks detect both human and computer errors. Some locate errors and some can be extended to overcome errors. Generally, any check that fails generates an output message in the error list.

THE RECOVERY SUBSYSTEM

The system will break down occasionally due to hardware and software failures so a recovery subsystem is necessary to:

— prepare for these failures;

— continue working using a fallback mode of operation when failures occur;

— recover from failures when they have been corrected.

Preparation usually involves procedures to ensure that information is not completely lost to the system when a crash occurs, even though it may be lost temporarily. Thus the system usually copies information at appropriate intervals of time. Typically, dumps of the database, input messages and before/after increments to the database will permit eventual file recovery. Sometimes, the database is duplexed or triplexed so that another version can be quickly activated if one fails. These dumps and copies can be regarded as part of the total database for a system.

Fallback implies that some parts of the system can run although other parts have failed. For example, the application subsystem works but the information retrieval subsystem does not, or 10% of a product file is blocked out but the rest is usable. Thus, fallback procedures divert input messages that cannot be processed and often store them in the database for recall when all is well. Graceful degradation means that fewer parts of the system run as more failures are progressively discovered (possibly by fault-finding procedures in the fallback subsystem itself). Ultimately, no parts can run and the system has failed totally.

Recovery needs procedures to reconstitute the database and reinstate delayed input messages. These put the system back on the air when hardware and software failures have been corrected by engineers and programmers. Recovery parameters are input messages, often invisible to the user, which control recovery by, for example, defining the break-point from which to recommence the system. Any output messages, for example warning a terminal operator that part of the database is not currently available, are usually regarded as part of the log.

The recovery subsystem can be large and complex for some real-time systems which are permanently on-line. Some batch-processing systems have few recovery procedures, for example a small monthly payroll system may contain no break-points for dumping and no fall-back procedures, because it is completely re-run after failure. Nowadays, standard systems software often provides many of the recovery procedures mentioned above.

THE MONITORING SUBSYSTEM

Each run of a system should be supervised to ensure that it has been operated according to plan, and to extract details which may assist current and future development of the system. The monitoring subsystem acts in much the same way that O and M and work study officers oversee human systems. Monitoring procedures produce a log which records both the sequence of operational events and useful statistics. Parameters are often input to control which statistics are output. The procedures include counting, summing, sampling, taking 'snap shots' of the system and can be programmed into the standard operating systems software, special hardware and software monitors, and the applications software.

Standard operating systems tend to log events that are mainly concerned with the operation of the system. These include the system's run time (usually the start, finish and hardware utilisation times of each program in the system), the sequence of events (particularly operator/operating system communications) and some warnings of inefficiencies in the physical system (such as the number of times records are retrieved from extended overflow areas in an indexed sequential file). This aspect of the log enables management to analyse computer usage, to re-run part of the system if necessary, and to detect some inefficiencies in computer utilisation. Some logs include computer usage costs to enable operations management to charge users for computer run time and resources.

Special monitors are often inserted into the system to capture performance date, usually on an ad hoc basis; they may be special hardware devices or systems software routines. Hardware monitors record the use of particular hardware components, for example the activity of a channel linking secondary and primary storage or the movements made by a disk arm. Software monitors do a similar job for parts of the computer software, for example the number of times a particular routine is obeyed and the computer time it uses might be recorded. These entries in the log enable programmers and systems designers to improve the efficiency of the system by tuning it.

Applications software generally makes little or no contribution to the log in practice, but can provide some of its most useful aspects, including the statistics discussed in later sections of this book (for example the number of each type of input message and output message, the hit ratios of files on the database and the frequencies of procedural decision exits). These data are more concerned with the overall traffic in a system and not only enable systems designers to significantly improve its efficiency but also help systems analysts to improve its effectiveness.

For example, too many amendments to the database may indicate that it is not staying abreast of events in the real world and that it should be updated more frequently.

The monitoring procedures in a financial accounting system are sometimes used by auditors to verify that the database is a true representation of the company's dealings and that the company is being correctly managed. For example, ten percent of customers might be sampled and their total gross value of outstanding orders might be summed to indicate if serious blockages are occurring in the order-delivery pipeline.

OTHER SUPPORTING SUBSYSTEMS

The previous six subsystems combine to form the whole operational system, but further procedures may be necessary to design, implement and maintain it, as follows.

Simulators are simplified models of the system which help to estimate its performance or to demonstrate it to users. An estimating simulator is usually programmed on the capacity and timing characteristics of computer hardware and systems software and on the flow of input and output messages and database records through the main systems procedures. It then calculates the probable use of computer storage and time; standard estimating packages are available at a price, for example SCERT (Systems and Computers Evaluation and Review Technique). A demonstration simulator is programmed on the basic parts of the eventual system to help explain what management are committing themselves to or to train clerks for their eventual use of the system, for example, terminal operators in some airline reservation systems are trained on these simulators before they 'fly' the real system.

Database creators are programs which convert the previous version of the database, possibly organised as manual records transcribed to some computer-readable medium (eg, punched cards), to the new version in readiness for the initial run of the system. Sometimes it is more convenient to write these special programs or to use standard software packages than to employ the normal insertion procedures. For example it could be very risky to assume a product file could be accurately set up by insertion facilities in the very first run of the new system; instead, it would probably be wiser to create this file before the new system commences to allow time for correcting any major errors.

Testing procedures are often necessary, particularly for complex real-time systems, to check that new programs or modifications to existing programs are correct. They can include procedures to:

— Prepare test data, particularly for exhaustive trials when large volumes and many combinations of input messages and database records are required.

— Trace tested procedures, as they are obeyed, so that the systems team can retrace the path of a trial, particularly if it fails. For example, an order processing system could include traps at strategic points to display the individual procedures that have been followed and their results during a trial run. Clearly, great care must be taken to ensure such traps do not interfere with the normal operation of a system otherwise a trial could work where a normal run does not.

— Compare test results against a correct version, for example to pinpoint differences between the test database output by a new system and that output by the previous system. If no correct version is available for comparison, then at least, post-mortem or dumping procedures must be available to print or display the contents of computer files. A new sales ledger system for an electrical company printed perfect statements, etc, from its first run and management were so delighted they decided further parallel runs between the old and new systems were unnecessary and therefore scrapped the old system. The next run was chaos because the database brought-forward on magnetic tapes from the first run was rubbish because the systems team had not exhaustively compared test results during trials. Correcting the new system took months resulting in a deterioration in cash-flow before any improvement.

Modular or structured programming techniques are now widely used to develop parts of a program independently of each other before they are combined. A 'test-harness' usually includes the above prepare, trace and compare facilities which can be applied to any part of a program that is undergoing trials in isolation.

These supporting subsystems may be developed for a particular application, but many are now available as general-purpose software, although they are expensive. Taking a wide view, much of systems software can be regarded as support for the operational system, for example loaders, assemblers, compilers and decision-table preprocessors.

Conclusion

This chapter has suggested the anatomy of a general system. The parts of a system should be identified before detailed fact-finding and specifying commences. The aim here is to provide a checklist of these parts, as follows:

1 The procedures of a system may be dissected into six subsystems:

application — to perform the user's business;

information retrieval — to interrogate the computer database;

database maintenance — to insert, amend and delete its contents;

control — to help detect, locate and correct errors in messages, records and procedures;

recovery — to prepare for, degrade during and recover from technical failures;

monitoring — to log operational events.

Many of these procedures are now incorporated into systems software and the user is often only aware of a small proportion of the remaining procedures. A continuing aim of systems software development is to standardise and simplify the housekeeping to reduce the inordinate amount of effort spent on it, hence the need for operating systems, file handlers, database management schemes, transaction monitors, application packages, programming utilities, and so on. The chief accountant of a civil engineering group refused to believe that his complex payroll system had required only minor modifications to the standard bureau system that he was using, but it was true.

2 The messages output by a system may be viewed as six types:

results — dictated by the application;

responses — displaying interrogated database information;

proof lists — showing changes that have been made to the database;

error lists — highlighting mistakes detected in information;

reconciliations — equating overall totals;

log — recording operational events.

Some systems may not include all these types, for example a library information retrieval system may write no results, its main output being responses, or a process control system may not use a database so responses and proof lists are not written.

3 The messages input by a system may also be viewed as six types:

transactions — to record real-world events which trigger results;

enquiries — to interrogate database information;

insertions — to set up new database records;

amendments — to change database elements;

deletions — to destroy old database records;

parameters — to trigger control, recovery and monitoring procedures.

Some types may be absent from a particular system, for example the middle four in a process control system without a database.

4 The database updated by a system records historic information which is needed by:

— the application subsystem to transform transactions into results;

— the information retrieval subsystem to transfer enquiries into results, including user profiles of interest;

— the control subsystem, particularly brought-forward totals and possibly constants used in data vetting;

— the recovery subsystem, particularly dumps and copies.

The database can also be viewed to include the programs of computer procedures, which will be maintained by software systems of assemblers and compilers, and various indexes, directories and pointers, which will help these programs to access the database efficiently.

Each system should be compared against such a checklist to ensure all parts have been included, and the general principles of previous sections have been fully considered.

Figure 2.1 illustrated the basic anatomy of a system to introduce the discussion of subsystems in this chapter but the diagram gives the false impression that the subsystems are independent. Figure 2.6 revises this view by incorporating the discussion, as follows:

1 All information can be controlled, recovered and monitored, therefore these three subsystems surround the other three and intercept all messages and database records, and they could be regarded as a super-control system.

2 The database is read by all subsystems and may be written by all except the information retrieval subsystem.

3 Messages are input and output as before.

Thus the arrow-heads in figure 2.6 indicate which subsystems read or write which information.

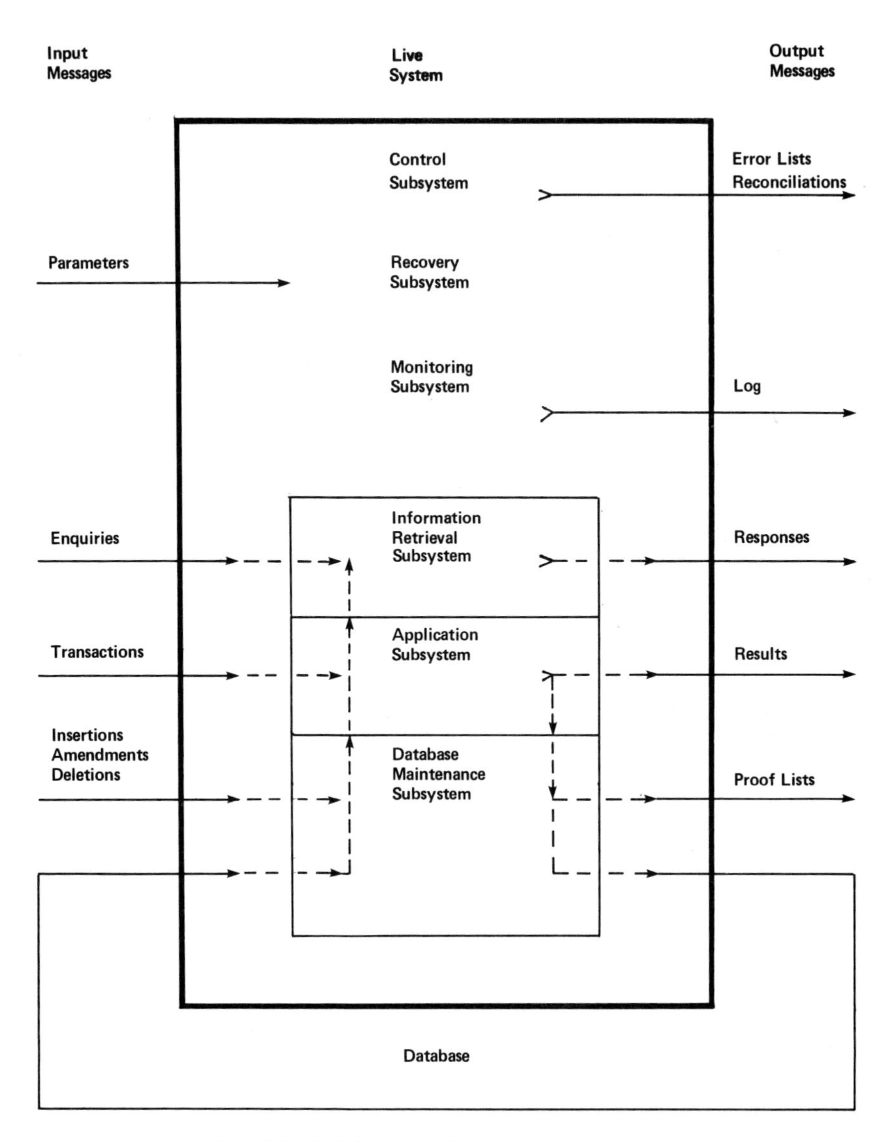

Figure 2.6 Revised anatomy of a system showing relationships between subsystems

3 Specifying a system

INTRODUCTION

This chapter develops some aspects of systems design. Several useful design techniques can be constructed to improve specifications.

The user often sees only a small proportion of the total system, which contains many types of messages, database records and procedures. Even small systems can require thousands of programmed statements; for example, a monthly-salaried payroll system needed five thousand lines of COBOL. Current experience indicates that the 'average programmer' produces about ten working statements per day at a cost of between £2 and £4 per statement; one nationalised industry estimates the cost of one programmer-day at £40 including overheads. Thus any system is likely to run into man-years of programming effort with costs being measured in tens of thousands of pounds.

These long times and high costs are not 'once-off' expenditures because after the system has been implemented it must be maintained for changes in user requirements. These may be due to central and local governments revising statutory procedures, national and international political pressures altering regulations, mergers, takeovers, market fluctuations and personnel changes affecting policies. Also, the system must be maintained to take advantage of technological progress. Probably most UK programmers are now maintaining systems rather than developing new ones.

Even the simplest modification can snowball into a major change. For example, if a system has been implemented and the user now requires a 'minor' revision to customer statements, various program changes may be necessary (see fig. 3.1.).

This chapter aims to suggest design techniques which can reduce large investment in systems implementation. These techniques are used

Subsystem	Output Messages	Input Messages	Database	Procedures
Application	Include Serial on Statements	–	Include Serial on Customer Record	1 Set all serials to zero at start of year 2 Add one to serial when Statement is prepared 3 Print Serial on Statement 4 Update Serial on Customer Record
Information Retrieval	Include Serial on Customer Responses	Include Serial on Customer Enquiries	–	Print Serial on Customer Response, if required
Database Maint-enance	Include Serial on Customer Proof Lists	Include Serial on Customer Amend-ments and possibly on Customer Insert-ions	–	Amend, and possibly Insert, Serial
Control	Include Serial on Customer Error List and possibly introduce another Running Recon-cialiation		1 Include Serial Controls (eg minimum and maximum values) 2 Include Brought-Forward Total of all Serials on Customer File, possibly	1 Check Serial when read or written via Messages or Database 2 Possibly, accum-ulate totals of all Customer Serials – Inserted – Amended – Deleted – Carried-Forward and Number of Statements 3 Print Serial on Customer Error List 4 Possibly, print Running Reconci-liation, replace Brought-Forward Total on Data-base by Carried-Forward Total & set this total to zero at start of year
Recovery	–	–	–	–
Monitoring	–	–	–	–

Figure 3.1 Possible ramifications of a 'minor' systems modification

when the system is specified in detail and the contents of each message, record and procedure are defined.

This chapter suggests techniques to specify the information elements in input and output messages and database records, and the detailed operations or statements in procedures.

SPECIFICATION METHODOLOGY

Specification methodology is the study of specification methods, a method is being a systematic way of doing a job, in this case detailed specification design. The contents of various types of messages, records and procedures must now be specified. The basic problem is where to start and where to go next, until the job has been completed. A logical sequence must be developed so that a systems analyst can concentrate on one aspect before proceeding to the next.

The classical method is illustrated in figure 3.2 and has been termed 'sucking back', 'derivation cascading', 'precedence analysis', and so on. First the information on output messages is specified because these are the user's reason for wanting the system. Next the procedures are defined, and all output information is traced back to deduce the required input messages. Finally the database is specified to provide historic information.

Thus the method contains four distinct steps with a proviso that each can successfully proceed to the next or return to the previous in case of failure. For example, it may not be possible to capture the input information required by a procedure, and an alternative procedure must be specified. Essentially, the output messages are analogous to finished products in a manufacturing system, the procedures are production processes, the input messages are raw materials and the database is the store of component parts.

There are many alternatives to the classical method but it remains popular because it is often the appropriate sequence, most systems being output-oriented. Also, it incorporates all the other methods, since it can be iterative. Refinements upon this method include:

— *total systems design,* where all systems in an organisation are planned together as one logical application before being split and implemented over a long time-scale as linked, physical applications;

— *top-down systems design,* where a (usually new) system is planned progressively at more detailed levels; for example, first at the level of files, then at the level of messages and records, and finally down to information elements;

Subsystem	Output messages	Input Messages	Database	Procedures
Application	A. Results	B. Transactions	C. Application Database	D. Application Procedures
Information Retrieval	E. User Responses (eg sales analyses) I. General Responses	F. User Enquiries (eg analyses triggers) J. General Enquiries	G. User Retrieval Database	H. User Retrieval Procedures K. General Retrieval Procedures
Database Maintenance	L. Proof Lists	M. Insertions Amendments and Deletions	–	N. Database Maintenance Procedures
Control	O. Error Lists S. Reconciliations	P. Control Parameters	Q. Control Database T. Brought-Forward Totals	R. Checking Procedures U. Reconciliation Procedures
Recovery	–	V. Recovery Parameters	–	W. Recovery Procedures
Monitoring	X. Log	Y. Monitoring Parameters	–	Z. Monitoring Procedures

Figure 3.2 Activities in specifying a system

— *bottom-up systems design,* where (usually existing) systems are bolted together to form a new one; for example, standard packages may be linked to yield an order processing system;

— *inside-out systems design,* where the basics of a system are designed before secondary elements are added.

There are various combinations of these refinements.

The main limitation to the classical method is that it assumes all the parts at one step can be treated similarly and before all those at the next. This is not possible and leads to numerous iterations within the method. A specification method is required which takes into account the various types of messages, records and procedures and incorporates their dependencies. The analysis and general principles of the previous chapter can be exploited here to develop such a method. The parts of a system are labelled as specification activities; which activities must precede others is determined and a precedence (eg PERT, CPA) network is constructed to yield feasible specification methods.

Figure 3.2 labels the parts of a system as specification activities and their precedences are established by tracing the discussion of Chapter 2 as follows:

1 Application subsystem

— the classical method can be applied here
∴A→D→B→C.

2 Information retrieval subsystem
— the special, user interrogation requirements can again be treated by the classical method
∴E→H→F→G.

— the general interrogation requirements may cover all database information but usually exclude the controls
∴C and G→I.

— thereafter, the classical method can again be applied
∴I→K→J.

3 Database maintenance subsystem
— all database information may be maintainable except for the controls, usually
∴C and G→M.

— thereafter, the input messages dictate the method
∴M→L→N.

4 Control subsystem
— error lists may cover all information except controls usually
∴ A,B,C,E,F,G.I,J,L,M,V,X and Y→O.

— thereafter, the classical method can again be applied
∴ O→R→P→Q.

— running reconciliations may cover any database information except the controls themselves
∴ C and G→S.

— thereafter, the classical method can again be applied
∴ S→U→T.

5 Recovery subsystem
— all information should be recoverable, but this subsystem is not usually concerned with the detailed contents of messages and records, thus recovery procedures and parameters can be specified independently
∴ W→V.

6 Monitoring subsystem
— all information and procedures may be monitored (other than the subsystem itself) but again the detailed contents of messages, records and procedures do not usually affect this subsystem so it can be specified independently
∴ X→Z→Y.

Figure 3.3 incorporates these dependencies into a network to show which parts of a system must first be specified down to detailed content before others can be. For example, insertions, amendments and deletions (M) should be defined before error lists (O).

This network is particularly useful when the specification work is shared among a team of systems analysts because it also indicates which activities can proceed simultaneously and independently. For example, one can work on the database maintenance subsystem (M→L→N) while another gets on with the control reconciliations subsystem (S→U→T). Thus project management can use the chart to establish efficient work-sharing, possibly even using PERT estimating techniques.

Alternatively, if only one systems analyst is designing the contents of messages, records and procedures, then the revised specification method of figure 3.4 can be followed. This incorporates the sequencing constraints of figure 3.3, includes the possible iterations, and is offered as an alternative to the classical method which ignores the anatomy of a system. The revised method structures the systems analyst to follow a systematic, step-by-step approach.

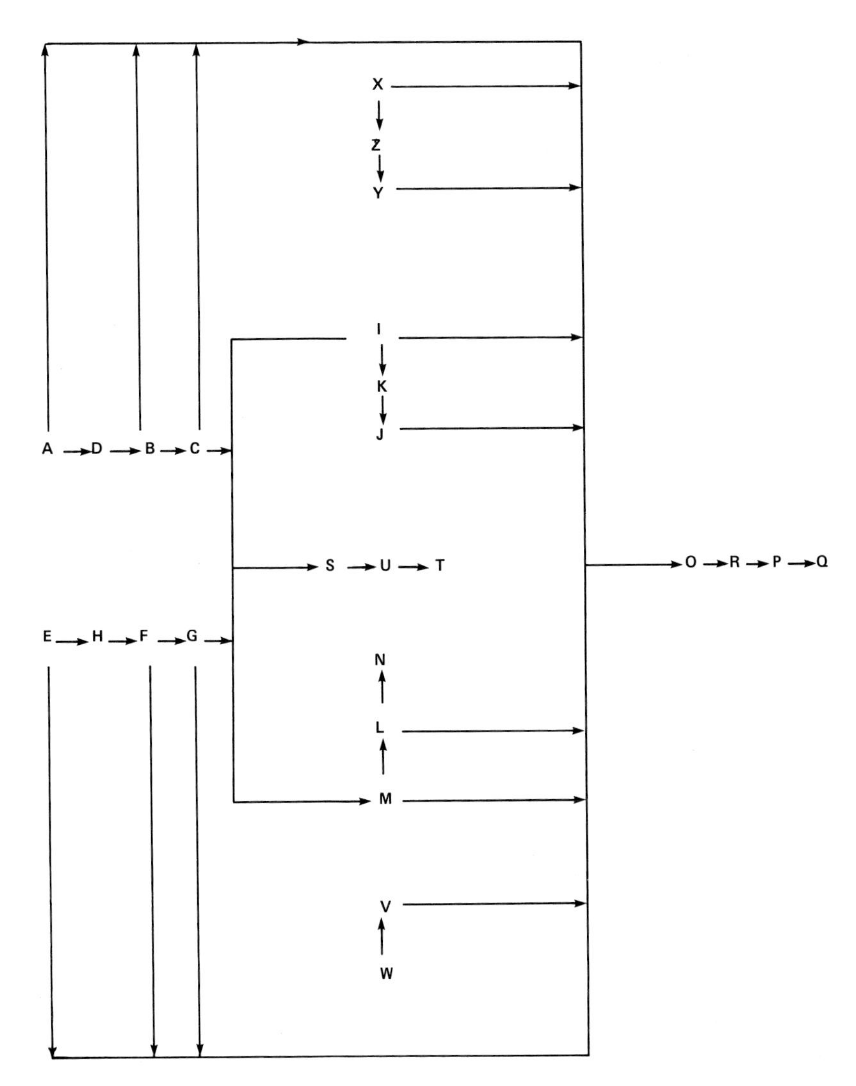

Figure 3.3 Precedence network of specification activities

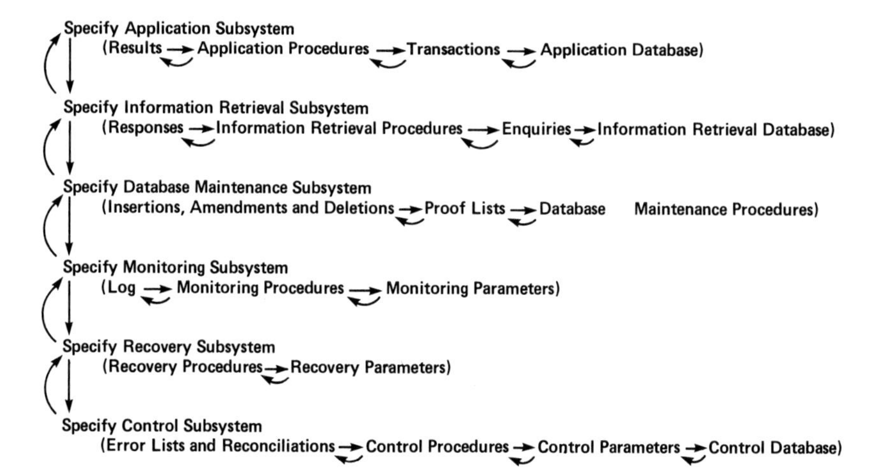

Figure 3.4 Revised specification method

MEETING OBJECTIVES

The specification method of the previous section should reduce the time and cost of defining a system in detail; specification techniques are now required to help it meet these objectives. During the initial defining of a system, a systems analyst can profoundly influence its eventual success. For example, if an inflexible version is specified, probably no amount of design and programming skill will subsequently make it flexible.

As an example consider the cash dissection procedure of many payroll systems: every employee is paid in cash, say to the nearest penny, and the procedure calculates the total number of each monetary (coin and note) denomination which must be drawn from a bank to make up the cash in all pay packets. Thus in British currency, the dissection must provide totals of the number of ten-, five-, one-pound notes, and fifty-, ten-, five-, two- and one-pence pieces.

SPECIFYING THE INFORMATION RETRIEVAL SUBSYSTEM

This subsystem can conveniently be regarded as two predefinable user interrogations such as regular sales analyses, and undefinable general interrogations such as ad hoc sales analyses.

SPECIFYING THE DATABASE MAINTENANCE SUBSYSTEM

Database amendment procedures can easily require thousands of program statements, particularly if each amendment message has its own routine for replacing the old value of database elements by their new values. Instead, it is often possible to specify one parameter-driven procedure which deals with all amendments, it simply selects each new value from an amendment message and stores it in the appropriate database element, 'appropriate' is defined by the parameters.

The particular approach can significantly impose on the manual or computer system but the general approach avoids this by simply amending the unit information (this automatically corrects the totals because they are summed from the unit information itself).

Insertions can also require many program statements to input the various messages and create new database records. This can often be avoided by viewing an insertion message logically as a collection of amendment messages which alter an 'empty' record. The insertion procedure simply sets up a record with its key number, and all other elements usually zero, and the amendment routines are then obeyed to complete the values of elements.

In this way, the database maintenance subsystem can be generalised by combining insertion and amendment procedures and by constructing a parameter-driven routine to amend any element.

SPECIFYING THE CONTROL SUBSYSTEM

This subsystem can conveniently be regarded as two, the checking procedures that produce error lists and the reconciliation procedures. Both can be generalised when specifying the elements in messages and records and the operations in procedures.

One parameter-driven routine can be specified to check all elements that are read from input messages and database and which are written to output messages and database, it can cover the format, radix, range, check-digit and existent code checks. The parameters for each element can include its format and radix, its minimum, maximum and intermediate choice of values and any check-digit formula. Similarly, another general routine can perform checks on messages and records, for example those on sequence, totals, completeness and comparison. Numerous standard packages which incorporate these controls are available, sometimes contained in data dictionary software.

The consistency check requires special treatment, as there are often complex relationships between information and these must be specified by the user. Decision table preprocessors are the most useful standard software which can be applied here. A database record should be checked for consistency whenever it is read or written and one routine should cover both jobs. Also, the same checks should be applied when a new record is inserted so the routine can also be used after an insertion message has been processed, and the same is true for an amended record so the routine can also be obeyed either after each amendment message has altered the record or after all have.

Reconciliation procedures accumulate numerous totals, usually hundreds in payroll systems, and present them as columnar accounts. Several running reconciliations may control the records of one database file and each requires a carried-forward total from the output records, an insertion total from the inserted records, an amendment total from the amended records and a deletion total from the deleted records. Also, it is usually wise to accumulate a brought-forward total from the input records so that it can be checked against the total held on the control record. A routine can be specified to add the appropriate elements from a record to one set of running totals and it can be obeyed when the record has been read, inserted, amended, deleted or written.

When the numerous reconciliations are finally written, thousands of program statements may be necessary to select the appropriate totals and arrange them in their correct places.

The control subsystem can be generalised by constructing parameter-driven procedures to check information and to prepare reconciliations. There are also opportunities to use the same procedure for several different purposes in order to simplify a detailed specification.

CONCLUSION

This chapter has suggested methods and techniques for detailed specification of a system. This will often reduce the times and costs of implementing and maintaining the system. The aim is to get the specification right before the system is constructed. A new systematic method has been distilled from the discussion in Chapter 2 to analyse specification activities. This should help management to plan work-sharing efficiently and to ensure that the right thing is documented at the right time. The method can be used within the various approaches to systems design, including top-down and bottom-up.

Next, several techniques have been indicated which generalise some of the patterns to be found in the information retrieval, database maintenance and control subsystems; these techniques are often also relevant to the remaining subsystems. Broadly, the aim is to simplify and rationalise systems so that they cost less to implement and maintain. In particular, flexibility can often be achieved by specifying parameter-driven procedures. Many of these techniques could be built into standard software to interrogate and maintain the database and to check and reconcile information, some already have been, but much scope still remains. Meanwhile, many organisations could benefit from generalising their own particular procedures so that each system makes full use of existing support programs.

Finally, the interface between systems analysts, designers and programmers can be hazy where the specification is concerned. This document, like the architect's blueprint and scale drawings, defines what is to be built in terms of input and output messages, database and procedures. Thus the systems team must co-operate over the specification in order to strike a satisfactory balance between the requirements of the user and the limitations of the computer. For example, specifying the contents of the information retrieval database involves many alternatives which must be weighed against conflicting objectives. The specification document must reflect how the system is to be built since it defines what is to be built. Thus the systems team must co-operate here and endeavour to get the specification right by meeting dominant objectives.

4 Specifying the data dictionary

INTRODUCTION

During the early 1960s, COBOL gradually became accepted as the major data processing language and it now dominates business programming. A COBOL program separates data from procedures (a data division specifies the information on messages and records and a procedure division states the operations that are performed on that information). Thus each program can describe its own data and this created great confusion in many systems.

Imagine a system requires twenty different programs, which are shared amongst ten programmers, and that five programs read and write the product file. Unless they are disciplined, each programmer will invent his own description of data and the product file may eventually be described in five different ways. For example, stock quantity may be called STOCK-QUANTITY, SQ, AMOUNT-AVAILABLE, NUMBER-ON-HAND and N-O-H. Thus the programmers each talk their own language and confuse each other and future generations who must maintain the system.

Many organisations soon recognised the problem and introduced disciplines to overcome the conflicts. All information on messages and records is described once, and each program must conform to this description. Often the specification included these standards and they were implemented using COBOL library facilities so that a program could simply call upon the data descriptions it needed. Nowadays, data-base management systems include data definition languages which are used to describe the logical and physical representation of information, independently from the programs that use it.

Data dictionaries are becoming popular for defining the identification, structure and characteristics of information and they can be regarded as the basis for specifications. The dictionary is often used throughout a systems project, thus it can help users to communicate with each other and with systems analysts during the early fact-finding stages until designers and programmers eventually use it to construct the system. Software is available to store and retrieve the data dictionary, generate COBOL data divisions and COBOL procedures to validate information.

NAMES

Element Names

Usually all information in a system is uniquely named so that procedures can identify the elements that they operate on; names are the variables that are algebraically manipulated by programs. The name of an element should not be confused with its values; QUANTITY-IN-STOCK refers to all products but the actual value of that element may vary between different products at a particular time. All information is named, this includes:

— elements on input and output messages and database records;

— transient elements which are strategically used to simplify procedures but are not read or written; these may be local to a single procedure or global to many procedures;

— constants or literals to the system (eg, tax rate). However, few elements retain constant value in practice so it is often wise to include them on the database instead of embedding them within programs.

Set Names

Many procedures will process sets of elements (eg, prepare and print invoice) so it is convenient to name them. Typically, these sets will be messages, records, files, etc, for example, ORDER, INVOICE, CUSTOMER-RECORD, OUTSTANDING-ORDERS-FILE.

Synonyms

In practice, names are usually chosen to be meaningful descriptions of what they represent, but different people may call the same thing by different names. Conflicts can arise when using names which are resolved either by insisting that everybody calls the same thing by the same unique name or by allowing synonyms, whereby many names may mean the same thing. In the latter, one name may be denoted as the commonest. The opposite case is that of homonyms where the same name is used to denote different pieces of information.

Meaningful names tend to become long and cumbersome (eg, EMPLOYEE-TAXABLE - GROSS-PAY -FOR - THIS - INCOME -TAX-YEAR-TO-DATE) so there is a natural tendency to adopt abbreviations (eg, EE-TAX-GROSS-YEAR or even EE-TGP-TY). Unfortunately these abbreviations may become meaningless to anyone other than their creators, but it seems sensible to allow shorthand when specifying a system or coding a program. Synonyms are a compromise because an abbreviation can be viewed as a synonym for the common, meaningful name and all that is needed is for handwritten abbreviations to be converted back to their names. People usually have to look up their abbreviations before using them so the simplest code numbers might even be sufficient (eg, 0001,0002, etc) and have the advantage of simplifying any supporting computer software.

States

Another convenient shorthand is to name the states that an element or set may take. For example if an employee database record contains a SEX-CODE which is zero for males and one for females, then there may be many procedures which test if an employee is male, by asking:

IF SEX-CODE = 1 GO TO . . .

This can become tedious and is less meaningful than asking:

IF MALE GO TO . . .

which uses a state of the SEX-CODE. Thus the states of an element may be named and the conditions for each state must then be defined. Similarly, sets may take named states. For example, CUSTOMER-ORDER-FOR-CATALOGUED-PRODUCTS might mean that all its order lines request catalogued products only so that the order must be dealt with immediately.

All these names, whether elements, sets, synonyms or states, contribute to the logic of a system and must be specified for designers and programmers.

LEVELS

Information elements and sets may be grouped into larger sets to form structures which are often trees but normally networks. The network includes input and output information in addition to the database. Such a structure reflects levels of information in the sense that a row contains elements or sets at the same depth. For example the fourth level can cover all input and output messages and database records.

Level numbers can be used to specify the contents of a set and to denote how deep it is. For example in COBOL a low-level number indicates a set that contains all following elements and sets with higher-level numbers. DATE-OF-ORDER contains only three elements (day, month, year) and DESCRIPTIVE-DATA contains at least five. Procedures can subsequently name any sets they operate on and the level numbering notation describes their contents. For example 'CREATE OUTSTANDING-ORDERS-DATA' means that values must be given to

PRODUCT-CODE, DATE-OF-ORDER, QUANTITY-OUTSTANDING, etc. In practice, level numbers tend to be assigned arbitrarily, thus CUSTOMER-RECORD may be given level 05 whereas PRODUCT-RECORD may be given 07. It is more sensible to use the same level number for all sets at the same depth, particularly if using some of the developing software which allows all sets at the same level to be analysed.

Levels are thus concerned with the logical structure of information and must be specified.

KEYS

Single Keys

In most systems, database information must be updated by application procedures, read by information retrieval procedures and changed by database maintenance procedures. Chapter 2 suggested that the retrieval and maintenance subsystems should cover all database elements, if only to ensure they are correct and to put right any errors. This means that every database element must be identified by a key so that one value of the key pin-points a particular value of the element, for example CUSTOMER-CODE is the key for DELIVERY-NAME-AND-ADDRESS so that code 123,456 possibly identifies the value of a customer's name and address. Thus if users want to interrogate or amend this customer's DELIVERY-NAME-AND-ADDRESS they must enter the CUSTOMER-CODE 123,456 on the appropriate input message.

A key is an information element that identifies an entity, which may be an object in the real world (eg, a shop or a product) or a concept in the real world (eg, a telephoned order or a date). Each occurrence of a particular entity, such as one shop out of all shops, is identified by a unique value of its key, such as CUSTOMER-CODE 123,456. Otherwise, keys can be regarded merely as elements that apart from this identification property, can be treated in precisely the same way as other elements. Of course, it can be logically argued that every element is a key. For example, QUANTITY-OUTSTANDING is a concept and it can take a range of unique values between 1 and 9,999 units, just as SALESMAN is an object taking values between 1 and 999, the only difference is that the first does not identify other elements but the second does (eg, SALES-TARGET). For example, an enquiry to retrieve the SALES-TARGET for SALESMAN 123 from the salesman file would produce one value, possibly £5,000, assuming he is on the file, and an enquiry to write the DATE-OF-ORDER for QUANTITY-OUTSTANDING 1,234 on the customer file could yield zero, one or many values depending on how many outstanding order lines are waiting for 1,234 units. SALESMAN uniquely identifies the element SALES-TARGET for that particular salesman but QUANTITY-OUTSTANDING does not uniquely identify any other element.

Thus a single key is one information element that uniquely identifies at least one other element on the database, but what of input and output messages? In practice, every output message is identified by keys so that the human system can relate its contents to the entities in the real world. For example a bad debt report must include the CUSTOMER-CODE so that management may know which particular customer is concerned. Keys are necessary in all information systems to identify elements with their appropriate entities. Similarly, all input messages should be identified by keys; most of them refer to particular database records. For example, a customer payment must include the CUSTOMER-CODE of that record to be credited. At first, it appears that some input messages, such as content-retrieval enquiries, do not need keys because they do not refer to particular records. For example, a request to find all customers with CREDIT-LIMIT exceeding £10,000 does not include a database key. However, since any input message may be found incorrect by the control subsystem it is wise to display a key, such as ENQUIRY-NUMBER, on the resulting error message to pinpoint the offender. Thus all information elements are usually identified by keys and these must be specified in the logic of a system.

Composite Keys

A composite key includes several elements instead of a single one. Lengthy identifiers often become unwieldy in practice because all elements must be included in such input messages as key-retrieval enquiries and amendments and instead, substitute identifiers are used to reduce human effort. For example, an INVOICE-NUMBER is generated by the computer system for each invoice and is subsequently used to uniquely identify it (instead of CUSTOMER-CODE/DAY-OF-INVOICE/MONTH-OF-INVOICE).

It can be argued that most keys are composite because time is always an element contributing to a key. CUSTOMER-CODE only identifies his PRIORITY if it is assumed the current database is being referred to. Here, time means now. If every version of the database that ever existed could be referred to, then the key must be TIME/CUSTOMER-CODE, because his priority could be amended over different versions. This argument yields the proposition that every element must have a key, even if it is merely time. For example, it may seem that an absolute fact such as the element DATE-LEO-WAS-FOUNDED must take only one value, 18-07-1940, and therefore needs no key. But if an incorrect value were set up in early databases and was eventually corrected, these differing values must be identified by time to differentiate between them. However, in practice in most systems time refers to the current situation, and if elements such as QUANTITY-OUTSTANDING may have several values, then a date (or some other key) is associated with each value.

Alternate Keys

Information elements and sets may be identified by more than one key because different users code the same entity in different ways. The previous section noted that an invoice could be identified by an INVOICE-NUMBER or by CUSTOMER-CODE/DAY-OF-INVOICE/MONTH-OF-INVOICE where the first key is strategically used to reduce human effort. Another example is an employee file shared between payroll and personnel systems where employees may be uniquely identified by their FACTORY-NUMBER/DEPARTMENT-NUMBER/CLOCK-NUMBER or by their NATIONAL-HEALTH-INSURANCE-NUMBER. Payroll transactions such as clock cards may include the first key but personnel enquiries may use the second.

Key Relations

In addition to identifying other elements, a key is itself an element which can be identified by another key. The key relations must be included in the specification so that a designer can plan the structure of the database and the access paths of each input message through the database into output messages. Simply, a key identifies an entity, for which information is recorded, and if one information set (eg, message, record) is related to another then it must contain its key to denote the relation. In practice, this is how most users think and also how manual systems were operated, before computers were invented.

Physically, these keys may eventually be replaced by pointers or some other computing technique, but this strategy is of no concern to the user and his specification of requirements. Further, he can ignore the various techniques for organising and accessing the database by simply assuming it is searched from beginning to end whenever its information is needed.

PICTURES

All information elements consist of character-strings with a maximum length and their characters may be numeric, alphabetic, and so on. Pictures are used to define the length and character set of every element. For example, CUSTOMER-CODE is six numeric digits, PRODUCT-DESCRIPTION is up to twenty alphabetic characters and DATE is two numeric, three alphabetic and two numeric characters (eg, 18JUL40). The picture of an element need not always be the same for all occurrences of that element, for example DELIVERY-NAME-AND-ADDRESS might be up to one hundred alpha-numeric characters for UK customers but two hundred for export customers. The maximum length of an element should include room for expansion to prevent its values overflowing the limit. A common rule-of-thumb for sterling amounts is to take the current highest value and add an extra digit to cope with inflation. For example, if the largest SELLING-PRICE is £87.50 now, then allow five decimal digits.

Pictures contribute to the logic of a system because they can indicate procedures which convert elements from one format to another and which are so obvious that they are not specified in detail. An input date might be defined with a picture of six numeric digits (180740) but its output version could be converted to two numeric, nine alphabetic and four numeric characters (18 JULY 1940). The conversion procedure must be programmed, including a table of twelve month names, but is considered to be so obvious that it is not worth specifying the detail. Pictures are also used for estimating how much computer storage will be required.

UNITS

Many numeric elements are not merely decimal numbers but measurements of some kind that require units of definition. QUANTITY-IN-STOCK might be measured in grammes/kilos or metres/centimetres or even by value of pounds/pence and its units should be specified. An element's units need not always be the same for all occurrences of that element, for example TRANSACTION-AMOUNT would vary its currency depending on the countries of export customers.

Units contribute to procedural logic because they indicate conversions between measurements. QUANTITY-DELIVERED might have units metres/centimetres (4 metres, 6 centimetres), SELLING-PRICE could be sterling per metre (£10.00 per metre) and the procedure for calculating the sterling GROSS-VALUE of an invoice line might simply be specified as QUANTITY-DELIVERED x SELLING-PRICE; the program must eventually extend this calculation into:

$$\text{GROSS VALUE} = \frac{(\text{QUANTITY-DELIVERED-FEET} \times 12) + \text{QUANTITY-DELIVERED-INCHES} \times \text{SELLING-PRICE}}{36}$$

Thus units help to complete some procedures and they also define radix checks in the control subsystem.

VALUES

Value Ranges

Elements take values, subject to the constraint of their pictures, which specify maximum lengths. The minimum and maximum values that each decimal element should not exceed must be defined so that the range checks of the control subsystem can be programmed. For example, QUANTITY-OUTSTANDING lies between 1 and 8,000. The minimum value of an amount is usually zero, but many systems have been caught out by supposedly positive values going negative. Any total that can be reduced by negative adjustments might legitimately become negative itself; the debt of a customer becomes negative if he overpays. In a payroll system, the total tax paid by an employee during the year should never be negative, but his tax this week could be a refund (ie, negative).

Therefore the total tax paid during the year in his current employment could be negative and even the total tax paid by all employees this week could be (eg, if the tax rate is drastically reduced). In some cases, an element may take values in discrete ranges and the minimum and maximum value of each should be specified.

Normal Values

The normal value of an element is the one that occurs most often. It is sometimes called the 'norm' or misnamed the 'average' although it rarely equals a half (minimum value plus maximum value). This statistic is very important to a designer because it can be used to estimate the size of variable length records of computer storage and to trace the most common paths through a system.

Actual Values

Sometimes the systems designer must be told all the values that an element may take, usually a key. For example if an algorithm (eg, prime divisor, digit selection) is being considered for LEO's product file then the actual values of PRODUCT-CODE may be analysed to find how many synonyms occur. This helps to estimate the time taken to find records on such a file.

Value Distributions

Sometimes a designer needs to know the number of occurrences of each value or range of values for some elements. The designer may be considering an inverted file for customers and will ask how many have top, medium and bottom priority or how many belong to each market area. These facts will help in designing the indexes and estimating file processing time.

Value Relations

Many relationships can occur between the values of elements. Payroll systems can yield hundreds of such value relations The consistency checks of the control subsystem cover these relations but they may also be needed when designing files.

It can be argued that these value relations will be specified in consistency checking procedures anyway so they need not be repeated in the data dictionary. However, some analysts and designers prefer to include them here to ensure their importance is not overlooked.

5 Specifying the messages

INTRODUCTION

This chapter discusses the facts to be specified for both input and output messages. There are six different types of input and output messages, but the same facts may be recorded for all of them. Sometimes, there are very strong similarities between input and output messages when:

— *turnaround documents* are written by the computer system to be read by the system later;

— *interactive conversations or dialogues* take place between man and machine. Here each interrogation triggers a chain of enquiries and responses which follows a predetermined pattern.

These similarities mean that all facts need not be documented laboriously for every input and output message. Instead one may simply refer to another. For example, the organisation of an input remittance advice is the same as specified for the output turnaround document.

CONTENTS

The contents of every message must be defined in terms of its information sets and elements, including the keys it contains. Since every message is usually named in the data dictionary, that will supply its contents via the level-numbering and naming facilities. Thus it will only be necessary to repeat the message name here as a reference to the data dictionary. A message is merely what we find convenient to term as such.

LOCATIONS

Every input message comes from a source location and every output message goes to a destination location. In typical batch-processing systems it is not necessary to document these locations as design facts because the interface between the user and the system is buffered by data preparation and assembly-disassembly sections. All input comes from one source and all output goes to one destination.

If teleprocessing or data transmission is to be used between user terminals and the computer system, then the communications network design is usually regarded as part (albeit specialised) of the technical system design problem. The location of every input and output terminal must be recorded so that the designer can plan the network. Also, any fact in this chapter may need to be specified for each location that generates a particular message because the facts may vary significantly between locations. This point is dealt with in each subsequent section.

If twenty market areas each contained a sales branch which took customer orders by telephone and then transmitted them directly to an on-line computer system, the designer must be told where these branches are geographically located in order to plan a communications network, in conjunction with the Post Office and terminal suppliers. This will link the branches directly to the computer and its costs depend on the distances involved. The designer must also know the traffic between each terminal location and the computer to make sure that no linking line is overloaded. Some branches may need faster transmission lines than others. The designer might consider a distributed system in which several computers, possibly at different places, share the workload. Further strategic locations may subsequently be introduced into the network by the designer.

REPLICATIONS

Replications of input and output messages are the numbers of times that each message is produced, or their numbers of copies. Output messages are probably always replicated, if only to keep a copy in the computer department to help resolve queries. Computer managers will want their own evidence when accused of losing results, passing poorly-printed documents or similar misoperations. Often, several copies of an output message are required. Sometimes, an output message may be copied hundreds of times. For example, the central computer at a bank's head office may log a message to all terminals in its six hundred branches that the database has crashed again and cannot be interrogated until further notice.

Control Replications

Input messages are usually read by a system only once, but some control subsystems include a comparison check, and read two versions of a message to check they agree. Sometimes three (or more) versions are read, and if they disagree, the majority verdict is assumed to be the correct one. This technique, often called 'any two from three', is common in teleprocessing systems, particularly where error-prone telegraphy is used.

Similarly, output messages are sometimes repeated to apply controls in the human system that receives them. A drug dose may be displayed twice on a hospital terminal so that a nurse can check before injecting a patient.

Recovery Replications

Some recovery subsystems rely on copies of all input and output messages as a final backup if all else fails. A real-time, nationalised industry system prints a log which can be used to recreate the database in the extreme case of all incremental dump files being lost to the system.

Replications by Locations

The number of times an input message is repeated may vary between source locations.

Similarly, the replications of output messages can vary between destination locations. A Birmingham manager may insist on ten copies of his bad-debtors report, because he delegates them to his salesmen, and the Manchester manager requires only one, because he takes the necessary action himself. Local procedures often differ when left to the manager's discretion.

Turnaround documents may not require that both input message and output message replications be defined. For example a lineprinted telephone bill may contain the output message twice for control purposes and this is documented. The re-input payment advice contains this same replication and there is little point in documenting it all again, a reference to the output message is sufficient. The number of replications of each message is not only needed for re-estimating but also for programming the logic of a system.

FREQUENCIES

Input messages are read by a system as and when they arrive in real-time, or at least as and when batches of them are collected and submitted together. Their frequencies of arrival must be documented in terms of time for each type of input message so that the designer can plan the logic for receiving them and estimate it. CUSTOMER-ORDERS are all presented to the system by 11.30 am on five days per week or they are submitted in real-time between 9.00 am and 5.00 pm during these five days. If the real-time system is international, as in airline seat reservations, the working day has no local bounds and is round-the-clock.

Essentially, output messages are always triggered by input messages and in most cases the output message frequency equals that of the input message. Output message frequencies must be documented in these cases and the delayed effects can pose severe problems for the systems designer.

Frequencies by Time

Message frequencies can vary over time. In the rag trade, CUSTOMER-ORDERS may be presented daily during peak seasons, due to the high work-load, but only weekly during slack trading months. Similarly, a travel agency booking system might output confirmations in real-time during the busy spring period but only daily during the preceding, slow moving autumn season.

Frequencies by Locations

Message frequencies can also vary between locations. A Birmingham manager might demand a particular sales analysis, such as the number and value of orders per salesman, every day whereas the Manchester manager only requests it each week.

All these frequencies are necessary for estimating such things as the number of computer hours per day but they can also contribute to the logic of a system. For example a current account banking system that reads credit and debit transactions in real-time would be markedly different if client statements were updated in real-time rather than if the updating was delayed until nightly batch-processing runs. Notice that the turnaround/response time between reading an input batch/message and writing its triggered output batch/message must also be defined for all cases under systems objectives (viz: timeliness). In many real-time systems, the response time is qualified by a required level of service (eg, 95% of airline seat reservations must be completed within ten seconds) or even by the availability of the database to avoid sharing problems (eg, personnel enquiries will not be serviced during the weekly two-hour payroll run, which updates the employee file).

OCCURRENCES

Normal Occurrences

The occurrences of messages, and their sets and elements, define the traffic that must be processed by the system. In the simplest case, the occurrence of a type of message is the quantity that is read/written with respect to its frequency: eg, 5,000 CUSTOMER-ORDERs read per day and 1,000 CUSTOMER-FALLING-SALES-STATISTICS written per month. In many cases the occurrence of an output message equals that of its triggering input message (forty key-retrieval enquiries generate forty responses, assuming their database records exist). If not, the output occurrences must be independently documented (forty content-retrieval enquiries may yield thousands of responses or none at all).

Occurrences are also relevant at levels below a message, particularly when it contains sub-messages. For example, a CUSTOMER-ORDER can contain five CUSTOMER-ORDER-LINEs. Similarly, occurrences can also be documented at higher levels than a message when it is contained in super-messages; fifty CUSTOMER-ORDERs might be contained in a batch.

Minimum Occurrences

The minimum number of times a message can occur must be recorded because, although it appears to be a statistic, it can significantly affect the logic of a system. In most cases, the minimum occurrence is zero which means the system must be capable of running without that type of message (even though the circumstances may be extreme). For example:

- — a payroll system must run without clock cards when there is a strike, to update database counters;
- — there may be no CUSTOMER-ORDERs due to a postal strike, but the system must process other messages.

Such 'exception' messages clearly include enquiries, amendments, insertions and deletions but some messages have a minimum occurrence of one, because the system cannot run without their influence. A run control message dictates vital parameters, such as the date, run number and indicators triggering particular procedures, and must be read before the system can continue.

Maximum Occurrences

The maximum number of times each message can occur is a useful fact to the systems designer when checking that response/turnaround times can be met in peak conditions. Most messages have a maximum occurrence greater than one, and the remainder equal to one, particularly the vital inputs mentioned in the previous section, their minimum, normal and maximum occurrences all equal one, unless control replications apply. The maximum occurrences of insertion messages usually occur in the special case when the system first starts to run, assuming this is how the initial files are created.

Occurrences by Time

Occurrences can vary over time and the designer may need their distributions for estimating computer resources in different periods of the year. Distributions are particularly important in complex real-time systems which can be sensitive to variations in message occurrences.

Occurrences by Time

The number of messages that can occur for a particular key value is a vital fact when programming the logic of a system, and one that is often overlooked. The intuitive answer is usually that an employee can only have one clock card submitted, but in practice several is often the case:

- — Can several CUSTOMER-ORDERs be read for a particular customer in the same day? If so, procedures must clearly state whether they are to be combined or not. Must separate INVOICES be printed and is discount calculated separately for each order or in total (possibly at a higher rate) for all orders, and so on? No doubt some customers will occasionally order twice on the same day if some products were forgotten on the first order or different departments are responsible for ordering different products; even a postal strike causes orders that were posted on different days to be delivered on the same day. Even if all customers guarantee never to order more than once per day,the system may itself delay an order causing it to be read on a day when the customer has submitted another one; for example, the first order was rejected by the computer because it had been read incorrectly and had failed control checks so it is resubmitted the following day.

— Can one CUSTOMER-ORDER contain more than one CUSTOMER-ORDER-LINE for a particular PRODUCT-NUMBER? If so, should they be kept separate or should a procedure combine them? A customer might validly submit such an order to, for example, receive different pack sizes.

— Can several insertions or amendments be read for the same key and what action should be taken if this happens?

Occurrences by Locations

Messages occurrences can vary between locations and are needed when designing teleprocessing networks.

Thus, occurrences range from occasionally useful statistics to vital facts that contribute to systems logic.

SEQUENCES

Sequences of input and output messages are the orders in which they hit the system or are generated by it, respectively.

Message Type Sequences

Input messages are usually read in arbitrary order because the computer can do any necessary sorting cheaply, quickly and reliably. However, such broadly *random ordering* often includes some sequences; all input messages of a particular type (eg, CUSTOMER-ORDER transactions) may be read together and separately from all other types of messages, thus the message types are in sequence.

Similarly, output messages are usually differentiated and sequenced by their types. Most batch-processing systems write error lists and logs as and when their messages occur, followed by reconciliations before all other outputs. (There may be little point in printing reams of results if reconciliations contain discrepancies and the system has to be corrected and run again.) Thereafter, each type of result would be written in turn.

Sometimes different types of input and output messages are mixed into a predefined sequence.

Message Sequences

Occasionally all input messages of the same type are submitted in strict sequence; payroll clock cards may be assembled in employee number order because each worker must be able to find his own in order to punch it on a clocking-in/out machine. The information in a particular input message can be in sequence even though the message itself is not.

Alternatively, most output messages are always written in strict sequences. In batch-processing systems, piles of these messages must be distributed to their recipients, eg PAYSLIPS are printed in FACTORY - NUMBER/DEPARTMENT - NUMBER/EMPLOYEE-NUMBER order so that they can easily be separated and sent to their appropriate factories and departments. In real-time systems, output messages are written in time sequences if nothing else. In all systems, copies of output messages are retained by the human system to help answer subsequent queries from the environment, so strict sequences are vital to help locate an output message that is the subject of a query.

Near Sequences

Input messages may be read in near sequence and if designers are told about this they might be able to choose an efficient sorting method which saves considerable computer time and costs.

Although near sequences are not vital facts contributing to systems logic, they are probably useful for choosing sorting procedures and estimating them.

Sequences by Locations

Sequences of messages can vary between locations. Once again, user requirements are a function of the particular user so the system must cater for each manager's variations, assuming they are reasonable in terms of cost and time, and these variations must be recorded.

Sequences by Replications

Message sequences can also vary between replications. Many local authority taxing systems print HOUSEHOLDER-TAX-DEMANDS in ADDRESS order to take advantage of cheaper postal rates and the treasurer's department has its TAX-DEMANDS-COPY printed in TAX-ASSESSMENT-CODE sequence to answer subsequent enquiries.

Turnaround documents may not require that both input and output message sequences be specified. An electricity-meter reader's round list for one day is printed in meter address sequence, which must be documented. The re-input messages, extended by actual meter readings, are in precisely the same order, and a simple reference back to the output messages avoids the extra documentation.

All these aspects of sequencing, apart from near sequences, must be documented because they influence control procedures, such as the sequence check and sorting logic.

ERROR RATES

Input messages are clearly prone to many varieties of manual, electromechanical and electronic transmission errors and the control subsystem tries to detect, locate and sometimes correct these. However, most input errors are corrected by the human system, guided by the computer's error lists, which means that the corrected messages must be read again, and so on. A public utility system inputs meter-readings and consistently rejects one tenth at the first attempt. These are manually corrected and reread to yield a similar proportion of errors which, being now 1% of all messages, are resubmitted the following day.

Similarly for output messages which are liable to be either misprinted or mistransmitted, the system must reproduce correct versions. This could mean for example, an extra hour per day on the computer's lineprinter. Error rates are very high in some teleprocessing systems and designers must include them in their estimates.

Error rates are often stated globally, a 5% overhead is added to all LEO's reading and 1% to all writing. This should be qualified by time, for example, does the 5% mean every day or that all input messages are incorrect on every twentieth day? The rates can be defined for different message types, eg 8% of transactions are incorrect but only 0.05% of enquiries are.

MEDIA

Input messages can be read from a rich variety of data capture media and devices, including:

— paper tape, sometimes as a by-product of accounting machines, edged-punched card equipment and data-loggers;

— punched cards, possibly via mark-sensing or port-a-punch equipment;

— magnetic tape/disk encoders;

— tags and badges;

— terminals, including teletypewriters, visual display units, audiocouplers, DATAPAD;

— optical mark recognition (OMR) documents;

— optical character recognition (OCR) documents;

— magnetic ink character recognition (MICR) documents, etc.

Occasionally, replications of an input message are assigned to different media, but it is more common for this to occur between different source locations.

Output messages can be written to several media, including:

— paper tape;

— punched cards;

— lineprinters;

— terminals, including audio-response techniques;

— microfilm (COM) and microfiche, etc.

Replications of an output message can be written to different media. For example a real-time system may write message to remote user terminals and print a hard copy log for the computer department. Again, different destinations can use different media for the same output message.

The medium must be specified for each message so that the designer can plan read/write logic and estimate its performance.

FORMATS

Given the media of every message, their precise formats must also be specified to complete read/write logic. The situation of every element in the message is defined, for example as punched card columns or line-printer positions, and the formats can vary by replications and locations.

Turnaround documents may not require that both input and output message formats be recorded. The Barclaycard system punched a card denoting each client's code and payment which was subsequently read when the payment had been made.

6 Specifying the database

INTRODUCTION

A database management system is simply standard software that extends the operating system, particularly its file handling procedures. Its advantages are in implementation and maintenance. Its disadvantages include the cost of the package and its demands on computer storage and run time. Numerous organisations have decided that a database management scheme is cost-effective or sufficiently prestigious for them, but many more await improvements.

Some companies have no use for the package because careful planning during systems analysis and design solved most of the problems that the package tries to overcome. For example, files are consolidated and given standard definitions, simple file handlers are used as a logical/physical interface between programs and data, generalised information retrieval procedures are specified, and random updating and data-sharing are avoided. A chemical company runs a scaled-down version of LEO on its own large computer with virtual storage, database management and several other aspects. A container company buys a few service bureau hours each week to run a similar system but with larger data volumes on a computer which has a minimal operating system. Although these two applications do similar jobs, their design philosophies are completely different. The first has hundreds of programs but the second uses less than ten and such consolidation often eliminates the need for database management systems.

Although the packages are undoubtedly useful in the right circumstances, perhaps they are being oversold and history is being distorted to fabricate evidence that the database concept is a major breakthrough in systems thinking. We hear that in the bad old days each application was built with its own files, which caused numerous problems of inefficiency, lack of consistency, and so on, whereas the new technology allows us to avoid all this by concentrating on the data independently from its applications. Figure 6.1 summarises the different approaches.

In the 1960s, some companies used the logical approach of integrating systems and certainly did not allow files to proliferate independently. However, database management systems have not always prompted this approach. On the contrary, they are merely physical aids to implementing a logical approach that is already established in many organisations.

A database is simply all historic information that is maintained by the computer system. It contains elements that are grouped into records and each record is identified by a unique key (so that it can be interrogated and maintained); records are grouped into files which combine into the database. Records and files are logical sets which are chosen at the convenience of the user and systems analyst.

Systems designers are responsible for organising the physical manifestation of the database, unless it already exists. They may restructure records and files for technical convenience.

CONTENTS

The contents of every database record must be defined in terms of its information sets and elements, including the keys it contains. Since every record is usually named in the data dictionary, that will supply its contents via the level-numbering and naming facilities. Thus it will only be necessary to repeat the record name here as a reference to the data dictionary.

It is sometimes argued that database contents can be strategically decided from merely the specification of input and output messages, the systems designer having sole responsibility for planning its contents. However this assumes that all messages are predefined so that their needs can be included in the database. This can be impracticable if, for example all LEO's ad hoc sales analyses must be specified. Instead, we may state that the database must contain all INVOICES for this and the last year. This is perfectly reasonable and natural.

The user and systems analyst dictate to the systems designer what database information will be sufficient to answer unforeseen demands, and this may be subsequently negotiated. Of course, the user and analyst may pretend they are completely ignorant of the database's existence and instead, fabricate output messages (eg an occasional print of all INVOICES of the past two years) to force information into its contents, but this seems rather artificial.

LOCATIONS

Every database record is stored on computer media, mainly magnetic at present, and this is geographically located somewhere. Most systems

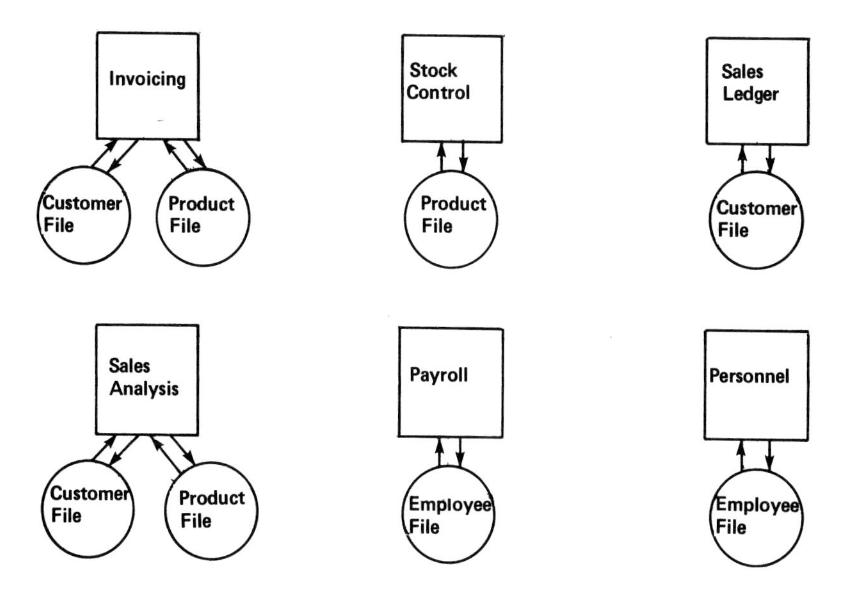

(a) Fragmented, disintegrated, patchwork approach of the bad old days

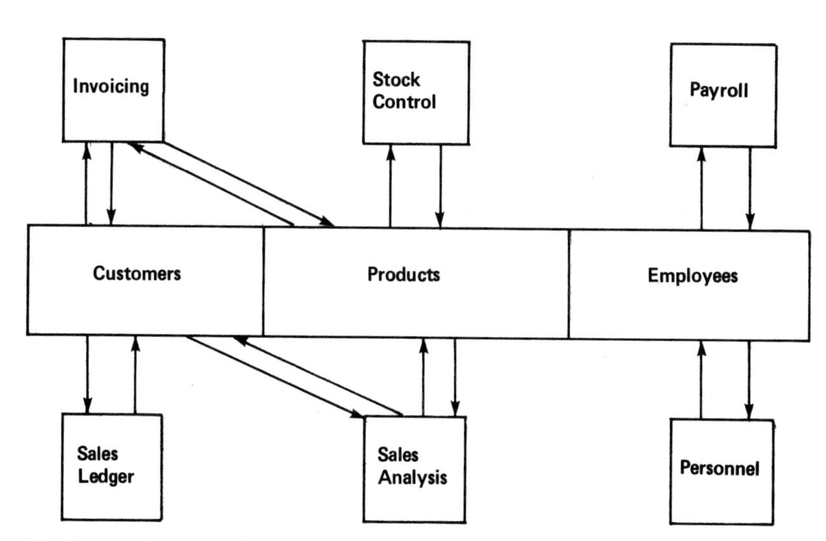

(b) Integrated, rational, database approach of the modern, golden age

Figure 6.1 Impact of database technology on systems thinking

use a single centralised computer and the database is situated with it. For example a bank's current accounts file is in the same room as its City of London computer (although security copies may be located elsewhere). Distributed systems contain several, perhaps many computers at different sites chosen for political, strategic and economic reasons. One of these considerations is where the database is to be located. For example, each autonomous company in a holding group may insist that its records are kept in its own head office so that it can manage its own affairs. The resulting hierarchical system might use a central machine at the group head office with a satellite at each company head office. Many organisations have shunned service bureaux because they refuse, often rightly, to lose absolute control over their data. Thus record locations must be documented if management have the power to dictate them, and these will usually suggest where the computers must be situated.

REPLICATIONS

Some of the early database management systems included the constraint that information must not be redundant, in the sense that only one version of each element should be stored.

There are however strong physical reasons why a designer might replicate information to improve the performance of systems. An engineering company, for example, uses two versions of its relatively static usage file, which defines the quantity of each resource needed to manufacture unit quantity of each product. One is stored in resource/product sequence and is used for product explosion, while the other is in product/resource sequence and is used for costing.

There are also strong logical reasons for replicating parts of the database. One is to control accuracy by using the comparison check; another is to quickly recover from failures. Such replications must be documented and, as with messages, they can vary between locations.

In the early days some organisations fragmented their systems and each manager owned his own applications and their data. LEO's chief accountant had his product file for valuation and the marketing manager had his for selling. Following integration the files are consolidated and shared between managers. One selling price is stored for each product and is used by both LEO managers for different purposes as they require the same values of prices. Or do they? Inevitably the chief accountant will want to increase the prices, to place a high value on stocks, whereas the marketing manager may want to decrease some, in order to shift these stocks. Sometimes managers will disagree on the values to be assigned to shared elements, so these must be distinguished from each other. Thus the consolidated file is reorganised to contain two selling prices for each product, one for each manager and so on, until in extreme cases the wheel turns full circle and each manager is happily restored with his own personal file. Systems analysts must take great care when deciding what database information can be shared and they must always remember that the power of a manager is heavily dependent on the information he controls.

FREQUENCIES

How frequently should database information be updated? Some argue that this is a technical strategy which is decided by the designer and is of no concern to the user. Given that we know the frequencies of input messages and their triggered output messages, including turnaround/ response times, then the database can be updated any time conveniently between the two.

Some real-time enquiry systems insist the response is up-to-date. For example, (how many seats are unfilled on a particular flight?'. If the reply gives old and therefore inaccurate information, an extra plane may have to be switched on for overbookings. Other enquiries can accept less current responses. The bank cited in the previous section asks 'what is the balance of a particular account?' and is not prepared to pay the price for anything more accurate than yesterday's closing balance. Other safe-guards are built into the total system to prevent large fraudulent over-drawals.

Thus the frequencies of updating database information are not merely strategic but also have logical importance to the user and must be specified. As with messages, these can vary over time and by locations.

OCCURRENCES

How large is the logical database? Some argue that this need not be documented in a specification because it merely reflects input and output messages so its size can be deduced from a full knowledge of the latter. For example if 10,000 product insertion messages are submitted to the very first run of a system to create that file, and if the subsequent numbers of product insertions and deletions are approximately equal, then there must be 10,000 product records on the database.

However, consider a possible statistics file which accumulates yearly sales totals for every customer and every product he has purchased during that year. How many customer/product records are there? At one extreme, customers might purchase the same five products every time they buy, so the answer is 200,000 x 5 = 1,000,000. Alternatively, they may buy five different products every time, and since the average number of orders per customer per year is 5,000 x 5 x 52÷200,000 = 6.5, the number of records is 200,000 x 5 x 6.5 = 6,500,000. But the true answer may lie somewhere between these extremes. The database often records message summaries over keys, particularly time, and their distributions must be recorded in order to estimate record occurrences.

It is much simpler to state the size of the database as occurrences, instead of deducing it from message occurrences. This avoids the need to record a potentially vast number of facts, such as distributions, and, since the user must know what is in the database because he inserts and deletes its records, why not simply ask him its size?

As with messages, the normal minimum (often zero to cover the special case of the very first runs, when the database is being initially created by insertions) and maximum occurrences may be specified, as well as their variations by time and location. Notice that only one occurrence of a database record type should occur for a particular key, otherwise it would not be uniquely indentifiable for access. Thus, unlike message occurrences, records should not have multiple occurrences for a key.

ACTIVITY STATISTICS

Systems designers should estimate database processing time to:

— ensure it is within the required turnaround/response time;

— check there is available computer time to cope with it;

— calculate its contribution to computing costs;

and thereby compare alternative designs. At least five measures should be documented to emphasise the database activity caused by input and output messages.

Hit Ratios

This defines the proportion of records in a file that are accessed, during a batch-processing run or period of real-time operation, as:

$$\text{Record hit ratio} = \frac{\text{Number of different records hit}}{\text{Number of records in file}}$$

being >0 and ≤ 1. This statistic can vary over:

— messages;

— time;

— location.

Hit Groups

This defines a highly-active area of a file as:

X (a high proportion) of hit records are contained in Y (a low proportion) of consecutive records of the file.

where $0<Y<X\leq 1$. For example 4/5 of all hits are in 1/5 of the file, which is Pareto's common situation. This statistic can vary over:

— messages;

— time (swimwear is popular in spring and football boots in autumn);

— location (in Wales the popular football boots are for rugby but elsewhere in Britain, they are for soccer).

Fan in/out Ratios

This defines the number of times each active record of a file is hit by input/output messages as:

$$\text{Fan in/out ratio} = \frac{\text{Number of input/output accesses to a file}}{\text{Number of different records hit}}$$

being $\geq$ 1. This statistic can vary over:

— messages;

— time (on average, 100 customers might demand the same product at the start of the season but only 2 at its end);

— location (Scotland's spending may be much less than London's, so fewer of its customers buy a particular product).

Volatilities

This measures the 'breathing' characteristics of a file in terms of the changes in its size and records. Some files expand, eg a medical research file of all live/dead cases. Others contract, eg a project control file of all activities still to be completed. Many remain static in size, eg LEO's customers, but these may change their records rapidly, eg the next hundred flights from London's Heathrow airport. It can be argued that:

$$\text{Size volatility} = \frac{\text{Number of insertions - Number of deletions}}{\text{Number of records in file}}$$

and:

$$\text{Content volatility} = \frac{\text{Number of insertions + Number of deletions}}{\text{2 x Number of records in file}}$$

where both are estimated over some period of time (usually per year), therefore they need not be specified. However, volatilities are so important in systems design that they are well worth stressing as separate facts in the specification. Again, different times and locations can change the volatilities of a file.

Overflow Patterns

Broadly this defines any distribution of records inserted into a file, particularly the point overflow situation where large numbers of new records are set up at the same spot of a logical file.

Systems designers will usually need to know activity statistics for each message and file, therefore they are classified as probable systems statistics. They may also demand some of their combinations depending on the design strategies being considered, for example:

— In a public utility meter-reading system, what are the combined hit groups for accounts paying their bills and those being updated by a new reading?

— In a current account banking system which captures transactions in real-time, but batch-updates the file several times a day, their record hit ratio might be 1/40 per hour. What is it every four hours, if that is the interval between the batch processes? It could be anywhere between 1/40 and 1/10.

These 'additions' of activity statistics may be highly significant to the designer but there is a potentially vast number that could be documented. For example, given ten types of input messages to a file, each with its own record hit ratio, then $2^{10} - 1 = 1023$ combinations are possible (of which the designer may only require a few, if any). Instead, these are best left to systems queries which are documented when requested by the designer.

CONCLUSION

At least eighteen facts can be recorded for each database record or file, of which nine contribute to the vital logic of a system. Remember that the activity statistics could generate a vast amount of work if all their possibilities were to be documented.

7 Specifying the procedures

INTRODUCTION

Procedures incorporate the decisions and actions necessary to transform input messages and database records into output messages and database records. Chapter 2 classified them into:

— application;

— information retrieval, including privacy restrictions;

— database maintenance, including ownership restrictions;

— control, including checks on batch totals, consistency, check-digits etc;

— recovery;

— monitoring.

The facts recorded in a specification are the same for all these types of procedures and cover their contents, sequences, frequencies and occurrences. Notice the similarity between these facts and those of the previous two chapters, arising because procedures eventually become computer programs which logically contribute to the database. Systems analysts record these facts for all logical procedures and they must be agreed by the user. Systems designers may subsequently negotiate changes to these logical procedures, for strategic reasons, and supplement them by numerous physical, computer housekeeping procedures.

In its simplest form, a procedure creates and destroys information, either as a value of an element or as an occurrence of a set of elements. For example, LEO's simple exponential formula calculates the SALES-FORECAST for each CATALOGUED-PRODUCT each week, thus a new value is created for this element and its previous value is destroyed. In practice, a single procedure in the logical specification is usually concerned with one element or set, but subsequently it may be split or combined with others to form physical program procedures.

Various documentation techniques are used to specify procedures, such as:

— formal narrative (figure 7.1 uses 'COBOL-ese' to define a possible discount calculation);

— algebraic formulae;

— flowcharts (figure 7.2 represents the discounting procedure);

— decision tables (figure 7.3 tabulates the calculation).

CONTENTS

A procedure contains statements which process input messages and database information into output messages and database information. Each statement contains, at least, a verb to define an operation and a set or element name as its operand; this name must be declared in the data dictionary.

Computer programming languages usually embody rich sets of verbs, which can define hundreds of different operations, but much of this richness is concerned with computing technique and is of little interest to the user. Logically, he merely needs to create (eg, calculate), test and destroy information and has no use for the computer's READ, MOVE and WRITE verbs, with all their ramifications of data representation, and only little use for looping. Only a few verbs are vital to a systems specification, plus their necessary naming conventions.

Names

The data dictionary (Chapter 4) named information sets and elements which are processed by the procedures, but a statement must be able to qualify a name by that of the set it is contained in. This is necessary when a particular set or element occurs in several sets of higher level.

Procedures need not themselves be named, if each one logically defines a single set or element, because the information name is sufficient. Thus, DISCOUNT-CALCULATION-PROCEDURE and its associated 'GO TO DISCOUNT-CALCULATION-PROCEDURE' are replaced by 'CALCULATE DISCOUNT', which names an element and thus its single procedure. Procedure names are only necessary when they are physically grouped to form a hierarchy of systems, such as program modules in programs in computer applications, and so on. Then procedure levels must also be defined. Individual statements may be named by labels if test operations branched to them.

A customer-order with gross-value less than £100 has zero discount.

If gross-value is not less than £5,000, then its discount-per cent is 10 if chain-store, else 8 if top priority, else 7.

If gross-value is less than £500, then its discount-per cent is 3 if top priority, else 2.

If gross-value is less than £2,000, then its discount-per cent is 4 if top priority, else 3.

If gross-value is less than £5,000, then its discount-per cent is 5 if top priority, else 4.

Discount = gross value x discount-per cent ÷ 100

Figure 7.1 Formal narrative for discounting procedure

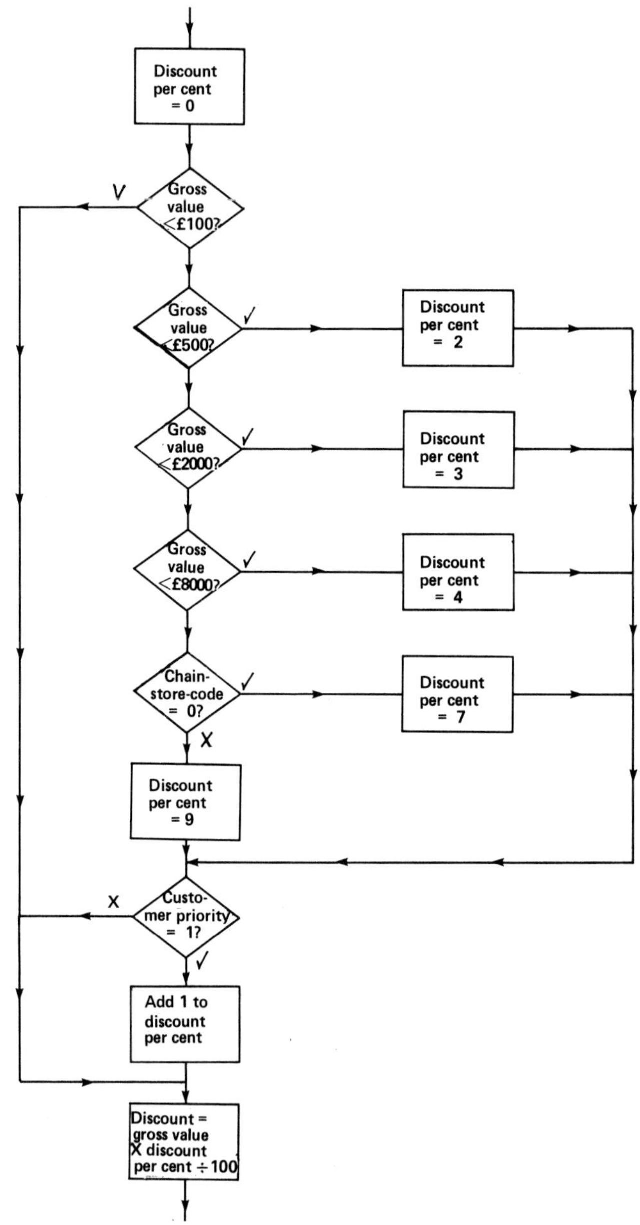

Figure 7.2 Flowchart for discounting procedure

		Rules									
	Decisions										
1.	Gross-value < £100?	✓	X	X	–	–	–	–	–	–	–
2.	Gross-value < £500?	–	✓	✓	X	X	–	–	–	–	–
3.	Gross-value < £2,000 ?	–	–	–	✓	✓	X	X	–	–	–
4.	Gross-Value < £5,000?	–	–	–	–	–	✓	✓	X	X	X
5.	Customer-priority = 1?	–	✓	X	✓	X	✓	X	✓	✓	X
6.	Chain-store code ≠ 0?	–	–	–	–	–	–	–	✓	X	–
	Actions										
1.	Discount-per cent =	0	3	2	4	3	5	4	10	8	7
2.	Discount = Gross-value X Discount per cent ÷ 100										

Figure 7.3 Decision table for discounting procedure

Create Verbs

Most action-taking verbs are of the form 'create a value for one information element by a calculation on the values of other elements', where the calculation includes mathematical and logical functions such as $+, -, x, \div, \Sigma, \Pi$, roots, powers, sine, cosine, tangent, AND, OR, NAND, NOR, and so on. The verbs can include rounding instructions, such as DOWN, OFF or UP, and each statement must avoid the usual pitfalls of dividing by zero, square-rooting a negative number or overflowing and underflowing the value range of the calculated element.

Keys and other elements can be created. A LEO INVOICE might include an INVOICE-NUMBER which is calculated as the previous value plus one (ie, a simple serial number). A mail-order house substitutes similar goods for those out of stock so a client requesting a red vacuum flask might receive a yellow one instead. Here, a procedure creates the delivery product code as being the order product code or if out of stock, its substitute, and the database lists the substitutes for each product.

Destroy Verbs

The old values of most elements are obliterated by the create verb that calculates their new values, 'CREATE INVOICE-NUMBER = INVOICE-NUMBER + 1' or 'ADD 1 to INVOICE-NUMBER' replace its old value by the new. Occasionally an element on an input message or database record must be destroyed, in the sense that it is no longer relevant to the system so delete it. For example, the mail-order house must be able to destroy a particular PRODUCT-NUMBER from the list of substitutes in a product database record.

Database records are sets of elements that must be destroyed whenever a deletion message instructs the system to do so. Similarly, input messages are sets of elements that must be destroyed, particularly if they fail control checks and so must be ignored by the system. A destruct verb is thus necessary for sets as well as elements, for example, 'DESTROY CUSTOMER-RECORD' or 'DESTROY CUSTOMER-ORDER-LINE'.

Test Verbs

Most decision-taking verbs are of the form 'test the value of one information element is less than, equal to or greater than that of another element', which may be a constant, sometimes zero, for example, 'TEST IF CUSTOMER-BALANCE IS GREATER THAN CUSTOMER-CREDIT-LIMIT?' or 'TEST IF GROSS-VALUE IS LESS THAN £100?'. Most tests yield a choice of subsequent action but some are unconditional changes in procedural sequence, such as 'GO TO CALCULATE DISCOUNT' or 'GO TO STATEMENT 3'.

Another form of decision is needed to test for the existence of a set, for example a database record or input message — 'TEST IF CUSTOMER-CODE IN CUSTOMER-RECORD = CUSTOMER-CODE

IN CUSTOMER-ORDER'. This verb is logically equivalent to searching files for matches on record keys and is vital to database maintenance procedures because:

— Before inserting a new record, a test is made that none already exists for its (unique) key. A procedure must define what action is taken when the test fails. For example:

1 Destroy the insertion or;

2 Destroy the insertion and the record or;

3 Destroy the record and create an inserted record or;

4 Create an inserted record with a different key; etc.

— Before amending a record, a test is made that it exists and a procedure must define the action if it does not. For example:

1 Destroy the amendment or;

2 Create a 'skeleton' record; etc.

Further, a test is made on whether several amendments occur for the same elements and keys, for example two price amendments for product 1234, which one, if either, should be effected?

— before deleting a record, a test is made that it exists and the deletion is destroyed if it does not. Sometimes deletion conditions are more comprehensive, for example only delete a:

1 CUSTOMER-RECORD if its CUSTOMER-BALANCE-OUTSTANDING is zero;

2 CUSTOMER-RECORD if it has no OUTSTANDING-ORDERS;

3 DEPARTMENT-RECORD if it contains no EMPLOYEE-RECORDS, etc.

Such alternatives complicate the database maintenance subsystem and have thwarted many standard software packages. The application subsystem may also include similar procedures, for example when no CUSTOMER-RECORD exists for a CUSTOMER-CODE IN CUSTOMER-ORDER. Usually the order is destroyed but some systems create a 'skeleton' record and process the order as normal.

The suggestion here is that only create, test and destroy verbs, with their variations, are necessary to specify logical procedures, including the triggers for when something is to be created/destroyed and the actions of how this is to be done.

FREQUENCIES

The frequency of operating a procedure can usually be deduced from the procedures of the messages it operates on. In real-time systems, an input message is usually converted into an output message within a few seconds' response time, so procedures used on the way have the same frequencies as these messages. For example, if five reservations are made per minute, the booking procedures are operated five times per minute. The same is true for most batch-processing systems in that input, procedural and output frequencies are the same, although staggered by the turnaround time. LEO's insertion messages, procedures and proof lists have daily frequency.

However, some output messages have frequencies different to their triggering input messages. LEO's CUSTOMER-ORDER-LINEs for special products are read daily but might only produce INVOICES weekly. Here the pricing procedure could be run daily, weekly or at some frequency between these extremes, such as twice weekly.

This example of LEO's special product orders illustrates the concept of implicit time. Inovices are produced at a predefined frequency after their orders have been input. Alternatively, time may be explicitly stated on the message itself. For example a PRODUCT-PRICE-AMENDMENT might notify the PRODUCT-CODE, NEW-PRICE and DATE-PRICE-TAKES-EFFECT, where the latter element dictates when the new value replaces the old. These delayed effect procedures influence database contents because it must store the message still in the pipeline at any time, for example the special product orders or all impending prices for each product. They can also influence procedures because these must state which version of a named element or set is relevant. For example, if an order was read before a price change but delivered after it (as with special products or outstanding orders), which price should apply?

Keys

The frequency of amending an element of one database record which is a key for another record, deserves emphasis because it can affect design strategy.

Sequence Keys

The frequency of amending the sequence key of a database record also deserves emphasis because it may cause the file to be reordered.

Notice that the designer may choose a sequence-key composed of several elements, eg the CUSTOMER-FILE could be ordered on PRIORITY/CHAIN-STORE-CODE/CUSTOMER-CODE or MARKET-AREA/SALESMAN/CUSTOMER-CODE, and so on, and the same problem arises if any of these elements changes its value. Instead of predicting all possible sequences, for every file, and documenting their delays (none of which might be used by the designer if he chooses direct or random access devices), these are best left as possible facts which are prompted by systems queries.

Sequence Key Series

Occasionally, the new values of amended sequence keys are always greater than their old values, and this condition can simplify the previous systems design problem as the additional file processing may be avoided. This fact might also be useful to a designer.

Sequence Key Cycles

This facility for amending sequence keys also enhances database maintenance procedures because appropriate actions must be specified if the new key is found to already exist on another record. Occasionally, this can occur legitimately if key 'swops' are allowed, so that key A changes to key B and vice-versa. This can extend to cycles where, for example, keys A_1, A_2, . . . , An change to keys A_2, A_3, . . . An, A_1. This fact also may be useful to a designer.

Strategic Delays

Systems designers may suggest many adjustments to specified frequencies, and introduce various delays, for the sake of technical strategy. Thus they will request many statistics, from a potentially vast set, to assist the designs.

Users tend to regard the system as a black box (see Chapter 1) in which all database information is readily available. This logical view corresponds to a physical, 'one-shot' design as typically used in real-time systems.

Thus in order to meet his objectives, a designer may distort the user's logical view of a procedure into some more satisfactory physical view. In doing so, he may need to know whether delays introduced are satisfactory or not.

OCCURRENCES

Given the frequency of running a procedure, it may also be necessary to know how many times it (or part of it) is obeyed, to enable occasional estimation of the computer time it takes. For example, in figure 7.3, the designer might need the occurrence of each rule, possibly expressed as a percentage of all occurrences. These are possibly useful statistics which are potentially vast and should be documented in reply to systems queries. The total occurrence of a procedure, being the number of times it is obeyed, is often (but not always) deducable from the occurrences of its messages and database information. As with messages and database records, the normal, minimum and maximum occurrences may be specified, as well as their variations by time, keys and locations.

SEQUENCES

Many documentation methods do not recognise that procedural sequences may be required by the user. Instead, they assume that all sorting is a strategic computer function to be employed at the discretion of the systems designer. However, many procedures must be run in a defined sequence to meet user requirements, just as most output messages must be sequenced.

A common example is that input messages must update database records in a sequence that matches the order of the real-world events which they represent. Often, new database records are inserted before amending the database, because a new record may incur an amendment, then transactions are applied, followed by enquiries and deletions. LEO's stock amendments should precede customer orders because they include stock-taking adjustments defining what is actually in stock, and not what the computer system may mistakenly think is in stock. Some banking systems insist customer credits be posted before debits whereas others take the opposite view and there can be significant differences between the effect of overdraft charges and stopped cheques!

Another example of a vital procedural sequence is the priority requirement that one information set (eg, input message, database record) has prior claim on a resource.

CONCLUSION

Figure 7.4 suggests that at least twenty facts can be recorded for each procedure, of which only six are logically vital. Most facts are possibly useful statistics which are best prompted by systems queries, particularly because a vast number of them are potentially recordable.

Replications are extremely important in systems design but they are more the concern of the designer than the user and systems analyst. For example, a system that must be highly accurate, such as an air-traffic control application, would require several computers to perform the procedures so that their results can be thoroughly checked. The designer must calculate service and reliability levels, based on technical repair and failure rates, of which users and systems analysts know little. Thus replications are dictated by accuracy and security objectives and not by facts recorded in the specification. Similarly, error rates are dictated by technical considerations which are more the concern of the designer than the user's specification although of course, the first must serve the second, and ensure accuracy and security levels are met.

Specification Fact	Systems Logic			Systems Statistics	
7.2 Contents	✓				
7.2.1 Names	✓				
7.2.2 Create verbs	✓				
7.2.3 Destroy verbs	✓				
7.2.4 Test verbs	✓				
7.3 Frequencies		✓			
7.3.1 Keys				✓	
7.3.2 Sequence keys					✓
7.3.3 Sequence key series					✓
7.3.4 Sequence key cycles					✓
7.3.5 Strategic delays					✓
7.3.6 Frequencies by time	✓				
7.3.7 Frequencies by locations	✓				
7.4 Normal occurrences					✓
7.4.1 Minimum occurrences					✓
7.4.2 Maximum occurrences					✓
7.4.3 Occurrences by time					✓
7.4.4 Occurrences by keys					✓
7.4.5 Occurrences by locations					✓
7.5 Sequences	✓				

Figure 7.4 Pro-forma of procedures facts

8 Checking the specification

INTRODUCTION

A specification documents the solution to the information problems of an organisation, for example LEO's business problems of management analyses, stock forecasting, cash flow, etc. This may well represent the most important aspect of the systems analyst's work because the users' needs are stated here. If these are wrong the eventual system may be worthless.

Many practical cases support this view. A petroleum company claimed to have reduced its £16,000,000 stores holding by over eighty percent by using computers, thereby eliminating five warehouses and five hundred employees. However, these savings were really made simply by scrapping obsolete parts.

The specification now defines a problem which is to be solved by technical systems design, usually yielding a computer-based system. Design and programming are pointless if the specification is either incomplete or inconsistent: a vital first step is to check the specification. Users, analysts and designers must all do this: users to ensure their information needs are met, analysts to co-ordinate their efforts (particularly if many of them have been documenting a large system for some months) and designers to verify that the statement of their problem is reasonable in terms of logic and statistics. This chapter suggests many checks that can be made on the syntactic completeness and consistency of a specification, particularly by a systems designer. These checking procedures are also an excellent discipline for helping designers comprehend the general and detailed logic of a system and to appreciate the workload it must satisfy.

OBJECTIVES

Chapter 1 suggested that the objectives and constraints to be met by a system should be stated at the beginning of its specification. It is important to check that they are. Figure 8.1 suggests objectives, which should be prominent in the specification. Notice that although some numbers are stated, much is left to human judgement so there is room for negotiation. In practice, users, analysts and designers are well advised to spend some time together discussing these objectives and refining them to indicate:

— the likely values of such benefits as faster response/turnaround times and higher accuracy;

— the likely changes that require flexibility and robustness;

— the expenditure flow of implementing and maintaining the system over its proposed lifetime, etc.

All these attempt to quantify objectives so that some form of cost/benefit analysis can give a clearer picture of the system's economy. Although the final go/no-go decision may be taken qualitatively, quantity can be an important influence.

Management consultants frequently emphasise that top management must play a significant role in every systems project; indeed its success depends on this. Many systems have foundered, cost small fortunes or crashed because higher management were not involved.

The objectives of a system may not be fully appreciated by middle managers who are concerned with the day-to-day running of an organisation. Often top managers are planning changes which have not yet filtered down, such as mergers, takeovers, rationalising products and customers, etc.

PARTS AND PRINCIPLES

Chapters 2 and 3 suggested a number of system parts and principles which should be seriously considered when specifying a system and its contents. These principles include:

— all database information should be open to interrogation by enquiries and responses;

— all database records should be insertable;

— all database elements should be amendable (possibly apart from keys);

— all database records should be deletable;

— all information read from input messages and database records and written to output messages and database records should be controlled by checks;

Objective	Importance	Notes
1 Efficiency	High	A large-scale, daily system so keep an eye on file sizes and computing costs!
2 Timeliness	High	Occasional delays can be suffered, provided they are only occasional!
3 Security	High	The customer file alone represents millions of pounds worth of debts owing to the company!
4 Accuracy	High	Occasional mistakes, such as wrong deliveries, can be tolerated!
5 Compatibility	Low	This is a stand-alone system!
6 Implementability	High	A ceiling of £100,000 on programming costs and it must be operational in one year!
7 Maintainability	High	The system must run for at least five years to pay for itself!
8 Flexibility	High	Ad hoc management analyses are vital!
9 Robustness	Low	No major disturbances, such as takeovers, are foreseen!
10 Portability	Average	A common hardware/software configuration is envisaged and the system will not be sold!
11 Acceptability	Low	Few design standards are imposed!
12 Economy	High	A ceiling of £200,000 per year on computer and maintenance programming costs.

Figure 8.1 Possible objectives

— control constants should be stored in the database;

— information retrieval procedures should be generalised;

— useful information should not be lost from the database (eg, by over-summarising it), therefore store unit information;

— amendment procedures should be generalised;

— insertion procedures should mainly use the amendment procedures;

— control (checking and reconciliation) procedures should be generalised.

FACTS

Nearly eighty facts may be specified, of which more than sixty are either necessarily or probably required. These facts should be recorded in the specification for the various parts of the system; otherwise, systems queries will be raised later and this can slow down the progress of the project (a work study of five computer installations showed that half of all programming time was wasted waiting for replies to systems queries). Omitted facts may not be spotted during implementation but will then come to light when the system is actually running. This wastes costly computer time, due to reruns, and often delays the system, eg a stockbroker's accounting system failed in almost half its runs during the first year of operation.

Check the level of confidence that may be placed on critical facts. For example, why build a costly, multi-threading system to process an optimistic forty reservations per minute if this is a wild guess and the number is nearer twenty (when a single-threading system might cope)? One expert claims virtually no forecast for one year and consequently is more than ten percent accurate. Many forecasts are wildly incorrect as UK government forecasting has demonstrated in recent years.

LOOSE-END ANALYSIS

All information on output messages must be derived by procedures from input messages and database information, so check that every element is created (unless it is given to the system). All information on input messages should be used by procedures, so check that it is, otherwise it may be redundant to the system, and similarly for database information.

These checks against missing definitions, redundancy and cycles find the loose-ends in a specification, and these must be tied up. Chapter 1 suggested that any system can be represented as a network, and one is extremely useful here for performing loose-end analysis, it is called a precedence network. Each node is an element (all elements are included) and each branch is the relation that its source element is used in the procedure that derives its destination element (all such relations are included). Thus:

means that element A is used to derive element B and so it must be derived first; in short, A precedes B.

A precedence network for elements can be transformed into one for sets, simply by grouping the elements into their set nodes.

Thus loose-end analysis checks the logical consistency of a system by tracing through its procedures to ensure that they derive all information not given on input messages, that all information not written on output messages is used and that no cycles cause problems.

ACCESS PATHS

Every element is uniquely identified by a key, so check that every procedure (whether it defines how or when) manipulates elements with related keys. For example:

'CALCULATE GROSS-PAY = HOURS-WORKED X SALES-PRICE'

is clearly a nonsense if the first two elements are identified by EMPLOYEE-NUMBER but the third is identified by a completely unrelated PRODUCT-CODE. An access path must be traced through every procedure by which a unique value can be accessed for each element through related keys. Essentially this is joining the sets of values for elements based on their keys.

CONSISTENCIES

Various facts which should be checked for consistency as follows:

— Check that the units of each derived element are consistent with those of its precedent elements. This is sometimes termed 'dimension analysis';

— Check that the minimum and maximum values of each derived element are not under/over-flowed;

— Check that the normal value of each derived element equals that calculated by its procedure on the normal values of its precedent elements;

— Check that the frequencies of messages, database records and procedures are consistent with each other;

— Check that the occurrences of messages, database records and procedures are consistent with each other;

— Check that the database activity parameters are consistent with the occurrences of messages and database records;

— Check that information named in the procedures has also been named in the data dictionary and that any ambiguities are resolved by qualified names.

These consistency checks can save a great deal of trouble later.

PITFALLS

Chapters 4, 5, 6 and 7 suggested there are many common pitfalls to avoid when specifying a system. These are well worth checking against and include:

— The minimum occurrences of messages, database records and procedures should normally be zero; If they are not, check that there are good reasons why.

— The maximum occurrences of insertion records must cover the initial creation of files, even to the extent of equalling the total number of database records for a 'big-bang' take-on.

— The number of messages that can occur for a particular key should normally be greater than one, if not, check that there are good reasons why, and ensure that procedures are specified to process these multiple occurrences.

— The number of database records that can occur for a particular key should not exceed one, so check against this.

— Avoid division by zero.

— Avoid square-rooting a negative number.

— Ensure all totals are set to zero, if and when necessary.

— Check procedures are specified to cope with amendments to elements that are relevant to delayed-effect messages.

Avoiding these pitfalls at the outset can save many crashes when the system is operational.

CONCLUSION

Is the amount of work involved in checking LEO's detailed specification pages really necessary? Although designers and programmers may be sorely tempted to dive straight into the technical deep-end, they do so at their peril. Always check the syntactic completeness and consistency of documentation before moving on to the next steps, otherwise errors will be found later and progress will be impossible. At each level, solve the design problem, document its solutions and then check this documentation, before proceeding to the next level. This saves a great deal of wasted effort later.

Computer software is being developed to assist in specification checking and thereby relieve the systems team of much of this routine drudgery. The facts are recorded in a higher-level systems language to become the input messages of a computer-aided design system. In essence, its specification checker is analogous to the data-validation procedures of any computer system.

9 Systems languages

INTRODUCTION

This chapter briefly considers a rapidly-developing area of informatics which is researching higher-level systems languages as a means of formally recording specifications. The user requirements can then be input to computer software which relieves the systems team of some of its drudgery and extends its capabilities.

Many languages are available, but only three have been chosen, for their original and pioneering work and for the wide publicity they have received. They are Information Algebra, Systematics and PSL. Each is discussed in terms of its objectives and main concepts.

These languages are all free-format, although forms have been developed for some users of the last two, but only PSL has supporting software. A fourth section surveys a number of other languages and their software to suggest that a computer-aided methodology of developing systems is becoming technically feasible.

INFORMATION ALGEBRA

Communications of the ACM, April 1962, included a report on 'An Information Algebra' by the CODASYL Committee (which was largely responsible for COBOL). The Language Structure Group (LSG) states its goal as 'to arrive at a proper structure for a machine-independent problem-defining language, at the systems level of data-processing'. The report does not claim to have achieved this goal but states: 'The LSG has not produced a user-orientated language for defining problems, nor has it specified the algorithm for translating information algebra statements into machine-language programs. In fact, these goals are probably unattainable in full generality.' The report is intended more as a discussion paper to 'stimulate others to think along similar lines' and to 'foster and guide the development of more universal programming languages'.

Information Algebra provides a mathematical shorthand for specifying procedures and the files they manipulate. However, the obscure terminology and notation of information algebra has understandably deterred many researchers in this area, often to the extent of their completely ignoring it. But the language is an interesting attempt to formalise specifications.

SYSTEMATICS

The word 'Systematics' suggests a methodical and scientific systems language.

The *Journal of the British Computer Society,* August 1966, included the paper 'Systematics — a non-programming language for designing and specifying commercial systems for computers'. This early paper describes 'a new language which provides tools and techniques for the systems analyst', which is 'an aid to problem solving' but 'is not provided with a compiler'. The objectives of systematics are 'to give precision to the building of information systems' models' and 'to be completely computer-independent'. These aims are very similar to those of Information Algebra.

This early version of Systematics contained three main features;

1 *Alternatives conditions* whereby the relationship between conditions and derivations (actions) were represented as a Boolean AND/OR matrix, subsequent issues of the *Journal* pointed out that this reinvented decision tables.

2 *Definition of qualities* whereby the states of an element could be defined as in COBOL.

3 *Classification of information* whereby an element could be classified as:

— permanent, updated, originated or destroyed;

— class hierarchy, which appeared to indicate identifiers;

— output, input or record (ie, database).

The paper includes a payroll example to illustrate these features in the form of a data dictionary and supporting procedures.

Systematics evolved over the next decade to culminate in the book *Systematics — a new approach to systems analysis* (McGraw-Hill, 1975). The language has the same broad objectives as before but they are restated as 'to help overcome the four traditional problems of the systems analyst, namely:

— to produce a statement of requirements, which is complete, unambiguous, short, and free from programming strategy;

— to obtain the user's agreement and, better, his understanding of what he has agreed;

— to communicate these requirements to programmers and minimise their chief time-waster — the systems query;

— to maintain an up-to-date description of what the company's computer system does.'

Systematics concentrates on when and what output is written, how it is derived from input and which access paths (or identification chains) link the one to the other. The emphasis is on tracing back the logic of a system to perform the loose-end analysis and access path checks.

Possibly, the most fascinating aspect of Systematics is its view of derivations being expressed ultimately in terms of input information, over all time. Then the database contents need not be specified because they are simply all messages that have ever been input to the system.

Systematics also includes some minor inconveniences which makes the language less elegant than it seems at first sight. One concerns deletions and their effects on the triggers of output messages.

Apart from an early, restricted experiment with compiler-compiler techniques, no software has been programmed to support the language and none is currently intended, thus systematics is again a communications aid. To return to its stated objectives, the language is not complete, therefore it will not minimise systems queries (but might help to reduce them). It demands not only an up-to-date description of what the system does now but also what it has ever done. However, it is short and free from programming strategy.

Whether users will ever become versed in Systematics remains to be seen. Although it is conceptually simpler than information algebra, Systematics is not easy to understand. If it fails in this respect, then specifications will not be documented in the language so it becomes an additional level of documentation which is independent from all the rest: the question then becomes 'is it worth it?'

PROBLEM STATEMENT LANGUAGE (PSL)

Datamation, August 15, 1971, included the paper 'Automation of system building'. This introduced the established ISDOS (Information System Design and Optimisation System) project. Figure 9.1 gives a very simple view of ISDOS as containing the following main software components:

1 PSA (Problem Statement Analyser) which reads user requirements specified in PSL and checks them for consistency;

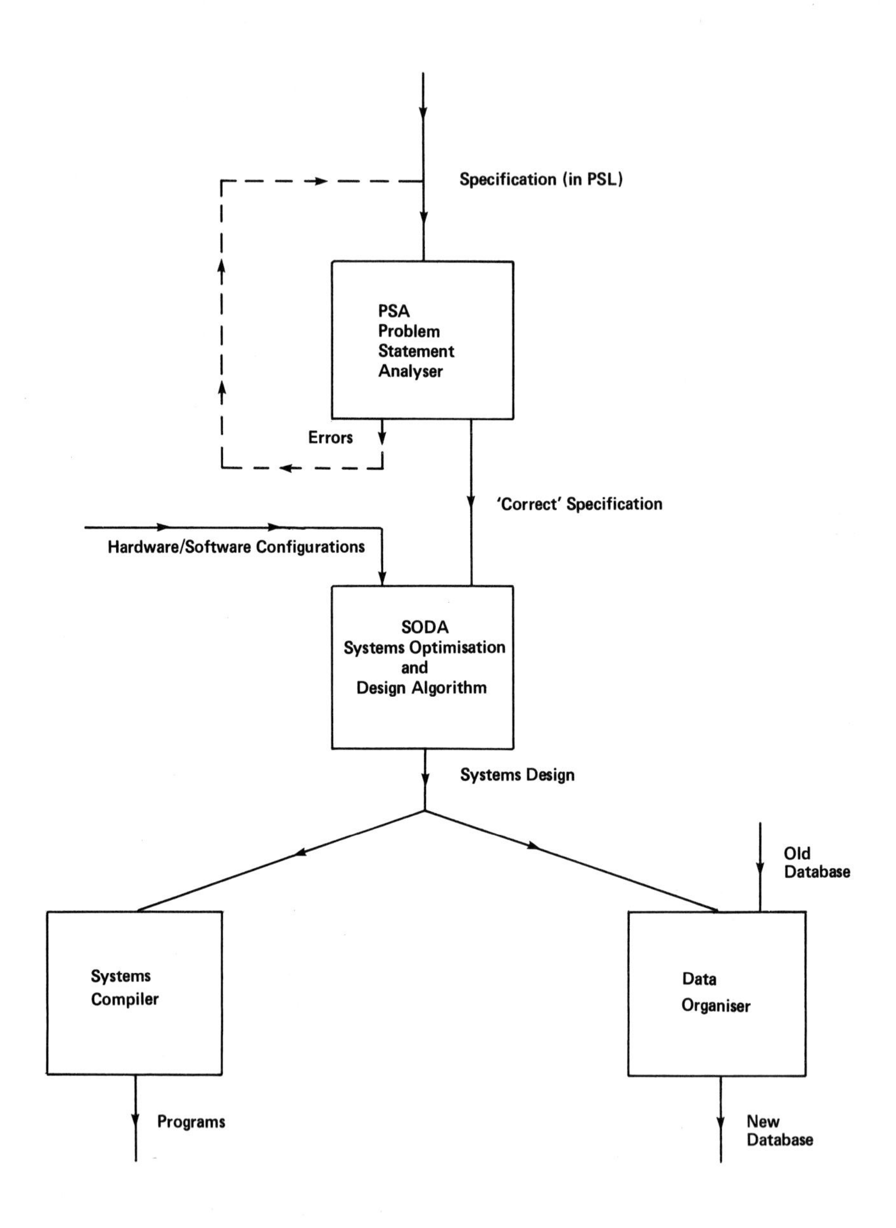

Figure 9.1 Outline of original ISDOS plan

2 SODA (Systems Optimisation and Design Algorithm) which automatically designs the computer system;

3 Systems Compiler which generates the programs by supplementing user procedures by computer housekeeping procedures;

4 Data Organiser which converts the old version of the database into its new, initial form.

'There has been a progression in which general-purpose programming languages have replaced assembly languages and general-purpose languages themselves have had to be augmented by database management systems to provide the framework for the programmer to communicate with the machine. In turn, the limitations of database management systems will be overcome through automation of the systems-building process.' Also, 'the central concept which makes possible the automation of design and construction is the separation of user requirements from decisions on how these requirements should be implemented. This philosophy is incorporated in the design of the Problem Statement Language'. Thus PSL has similar objectives to information algebra and systematics, plus full supporting software to design and implement a system.

PSL has evolved over the intervening years into a formal language, as specified by D Teichroew and M Bastarache in *PSL User's Manual,* ISDOS Working Paper No 98, March 1975. Companion papers describe PSL as 'a language for describing systems', which 'is *not* a (procedural) programming language', that improves the quality of documentation because its 'preciseness, consistency and completeness is increased'. Further, 'the objective of PSL is to be able to express in syntactically analysable form (ie, formally) as much of the information which commonly appears in system definition reports, (ie, specifications), as possible'.

PSL is not complete: it does not provide formal facilities for recording all facts. For example the procedures that derive elements cannot be specified formally. However, its generous description, attribute and memo statements allow any fact to be specified informally as comments. In any case, PSL is possibly the most complete systems language that is currently available.

PSL is supported by PSA software which progressively inputs user requirements during the course of systems analysis, checks them, and outputs reports to be included in the final specification and to assist subsequent systems design.

The following claims are included in the ISDOS literature:

— 'The cost of documenting specifications for a proposed system using PSL/PSA is approximately equal to the cost of carrying out the documentation manually.' Alternatively, it could be argued that more documentation is necessary, particularly because most relationships must be doubly-specified in PSL.

— 'The cost of typing manual documentation is roughly equal to the cost of entering PSL statements into the computer.' This is fair comment.

— 'The computer cost of using PSA is roughly equal to the cost of analyst time in carrying out the analyses manually.' However, this is not substantiated by any evidence regarding computer processing time and cost, quality and cost of systems analysts, etc.

— 'The major benefits are in the areas of improved 'quality' of documentation and reduced cost of implementation and maintenance.' The first point seems reasonable because of consistency checking and centralised co-ordination of the specification, but the second again needs substantiation.

— 'The co-ordination among analysts is greatly simplified since each can work independently and still have consistent systems specifications.' This is fair comment.

— 'The use of PSL/PSA during the specification stage greatly reduces the number of errors which will have to be corrected later.' This depends on the quality of the systems team it is being pitted against, but it may often be true and is an important benefit.

These are qualitative opinions that beg for some controlled experiments to help clarify the situation. At present the case for PSL/PSA is not clear but it could certainly be improved, by enhancing PSL.

Finally, the early ISDOS project was plagued by a series of exaggerated claims which oversold the research and reduced its credibility to many practitioners. These claims included 'complete systems language', 'automatic design' and 'automatic program generation'. The current ISDOS project at Michigan University, has clearly lowered its initially soaring level of ambition to concentrate on a single important problem area of informatics and has made an important contribution towards its solution. The early claims are now replaced by such refreshingly practical statements as 'completeness can never be fully guaranteed' and 'design is essentially a creative process and cannot be automated'. The future of ISDOS should be a promising one.

OTHER LANGUAGES

A brief glance through the literature shows the number of attempts over the past decade to create a higher-level systems language. They range from limited attempts at improving particular aspects of systems definition (eg, data dictionaries, decision tables) through to comprehensive schemes for general documentation (eg, natural language processors). Some aim for highly-stylised notations to condense documentation, similar to the confusing Information Algebra, whereas others are more intelligible extensions of COBOL-like languages, such as PSL.

Whatever the scope and form of a particular systems language, it is only a device for capturing the specification by computer. Then the main question is, 'what can the computer do with this documentation?'. Following, are a few notable examples:

DATAFLOW–An early, but now defunct, attempt by The National Computing Centre, UK, to define a language DATAWRITE and build a supporting consistency checker DATAFLOW (similar to PSL/PSA). Hearsay suggests that the language was unacceptable and the software took hours of computer time to process loose-end analysis.

Data Dictionaries — Probably more than a hundred data dictionary schemes have now been developed which record some facts and:

— check their consistency;

— retrieve them on request;

— generate COBOL data divisions;

— generate control checking procedures (eg, data validation programs), (or at least some of these).

CASCADE — A Solvberg of Trondheim University, Norway, is developing CASCADE (Computer-Aided Systems Construction and Documentation Environment) which builds on the pioneering, but largely impractical, work of B Langefors in *Theoretical Analysis of Information Systems.* The project uses a systems language to progressively document user requirements, and supporting software analyses information precedences and checks consistency.

LEGOL — R Stamper of LSE, UK, is developing LEGOL (a legally-oriented language) which records statutory law such as taxation, unemployment allowances, and so on. Typically these procedures are highly complex but have the advantage of being (reasonably) well documented and so make an excellent test-bed for any systems language trying to raise the level of specifications. The supporting software is based on relational database and macroprocessor technology and not only builds a static model for consistency checking but also a dynamic model which operates the procedures. Thus, when procedures are specified, values can be given to input information and the software derives the resulting output information. Procedures can be checked when they are defined and these verified procedures can subsequently be checked against their programmed versions, by generating test input and output.

SIMULATORS — SCERT (Systems and Computers Evaluation and Review Technique) and CASE (Computer-Aided System Evaluation) are examples of proprietary software packages which estimate the performance of a physical system on a variety of computer configurations. Each uses a language to specify the structure (programs and files) and characteristics (volumes and traffic) of a physical information system, and each contains the costs, times and capacities of the hardware and software provided by a wide range of computer manufacturers. Their estimates have been remarkably accurate for many organisations and applications.

MODEL — N Prynes of Pennsylvania University, USA is developing MODEL (Module Description Language) and a supporting processor which attempts to automate the design of a program (not a complete system). User procedures are interactively coded in MODEL; a precedence network is automatically constructed to determine the sequence in which individual procedures can be obeyed; and a PL/1 program is generated. Thus the aim is to generate the control module of a program from information precedences, in addition to applying the usual consistency checks and assisting the user to state his requirements. requirements.

PROTOSYSTEM — G Ruth of MIT, USA aims to automate systems design and programming, and the first version of his Automatic Programming System Prototype apparently takes some design decisions for limited systems (eg, organisation of batch-processing systems to minimise file accessing). The research continues. Several other USA Universities are working similarly by developing basic software that does part of the job for restricted environments, and then enhancing it gradually in the hope that it will ultimately become fully comprehensive.

SODA — J Nunamaker of Arizona University, USA, continues to develop the original ISDOS scheme of figure 10.15, particularly the automatic design aspect, SODA. Earlier versions of SODA were criticised for limited objectives (eg, minimise computer processing time); for incomplete PSL; for not being automatic (despite the claims that they were); and for omitting most design decisions, alternatives and techniques. It is now claimed that the latest version of SODA overcomes all these criticisms and operates satisfactorily.

AUTOMATIC PROGRAMMING — P Goldberg is leading a large research team at IBM, Yorktown Heights, USA which aims to automate business systems design and programming. Users state their requirements in BDL (Business Definition Language), via a two-dimensional visual display unit, in progressive levels of detail using the 'top-down' approach. Then preprogrammed fragments are automatically selected and composed into a computer system — these fragments 'represent all of the common combinations of performing a given business function'. The research is still in its early days of answering fundamental questions, such as 'how can the computer best extract requirements from a user?, but some interesting papers have already resulted from the project.

These projects are merely a few from the many researching and developing aspects of systems languages and their supporting software.

In the longer term, the key question is whether or not design will be totally automated. If so, the systems team will be released to program the system (eg, to specify the user's requirements in a systems language) and the computer will be left to program itself. Current hardware and software technology is not always as powerful as its manufacturers would lead us to believe. For example, there are many cases of badly-designed systems taking days to run which once corrected, take only minutes; ingenuity is often still required to design an efficient system.

CONCLUSION

Apparently no existing systems language reaches fifty percent coverage of specification facts; Information Algebra, Systematics and PSL are approximately twenty, fifteen and thirty percent 'complete', respectively. However, further research should correct this. Perhaps more disappointing is their complexity. Strange notations and extensive vocabularies can cause confusion, require much training, and deter potential users. Remembering that these users may not all be systems people suggests a prime objective for any systems language as simplicity. The documentation standards of the previous chapter exemplify this. Thus language designers might learn much from these standards, which typically take less than one day's training to understand (whereas each of the languages may take weeks to comprehend). Probably the answer lies in a compromise between the extremes of fixed-format and free-format so that standard forms cover those facts that suit them, and the remaining facts are covered by supporting statements.

Hopefully a comprehensive and comprehensible systems language will be invented and supporting software will be developed along the lines of a computer-aided methodology.